Faith Lutheran Church
Eight Sherwood Road at York
Cockeysville, Maryland 21030

**experiments
with everyday objects**

KEVIN GOLDSTEIN-JACKSON is a writer and television producer of children's programming in England.

NORMAN RUDNICK is a physicist and science editor.

RONALD HYMAN is Professor of Education at Rutgers University.

experiments with everyday objects

Science Activities
for Children, Parents, and Teachers

Kevin Goldstein-Jackson

Norman Rudnick

Ronald Hyman

Illustrations by Jonathan Harvey

A SPECTRUM BOOK

Prentice-Hall, Inc., *Englewood Cliffs, New Jersey 07632*

Library of Congress Cataloging in Publication Data

GOLDSTEIN-JACKSON, KEVIN (date)
   Experiments with everyday objects.

   (A Spectrum Book)
   1. Science—Experiments.   2. Science—Experiments
—Juvenile literature.   I. Rudnick, Norman, joint
author.   II. Hyman, Ronald T., joint author.   III. Ti-
tle.
Q164.G58   502'.8   77-13232
ISBN 0-13-295287-4
ISBN 0-13-295279-3 (pbk.)

Published by arrangement with Souvenir Press, Ltd., London.

A Spectrum Book

10   9   8   7   6   5   4   3   2   1

Printed in the United States of America

PRENTICE- HALL INTERNATIONAL, INC., *London*

PRENTICE-HALL OF AUSTRALIA PTY. LIMITED, *Sydney*

PRENTICE-HALL OF CANADA, LTD., *Toronto*

PRENTICE-HALL OF INDIA PRIVATE LIMITED, *New Delhi*

PRENTICE-HALL OF JAPAN, INC., *Tokyo*

PRENTICE-HALL OF SOUTHEAST ASIA PTE. LTD., *Singapore*

WHITEHALL BOOKS LIMITED, *Wellington, New Zealand*

The expansion of a child's mind
can be a beautiful growth.

—SYLVIA ASHTON-WARNER

contents

buoyancy

surface tension

mechanics

chemistry, colors, and candles

electricity and magnetism

preface

We have designed this book to involve you in the fun of science. We want science to come alive for you as you do the experiments that follow. There is little point in just *telling* you about air and water and other things in our world. On the other hand, there is much sense in having you *work directly* with the things in the world around you —to observe what happens, to understand why certain events occur, and to have fun while learning. Our hope is to arouse your scientific curiosity and imagination.

As you skim through these pages, you will quickly notice what is in store for you—a wealth of experiments to attract you ·to "do science." The experiments have been divided into seven groups to illustrate larger scientific ideas, and each experiment has several key features. *First,* the materials you need to do the experiment are mostly common household or school items that are readily available. You need not run to your local research laboratory to borrow them or to your local supply house to buy them. They are not expensive or complicated pieces of equipment. Your home or classroom or youth center can become your own personal laboratory. We list all this equipment clearly for you at the beginning of each experiment so you can prepare yourself and create your own "lab."

Second, each experiment that you do is simple and observable with your own eyes. By actually working with the materials and performing the experiment as directed, you will get the feel of being a scientist in your own laboratory. The key here is for you to follow the directions, do the experiments, and observe what happens. The experiments are harmless—provided, of course, that you are careful when using lighted candles, matches, and other materials. The fun and the learning begin when you start doing the experiments. Do them several times, if necessary, to get to know them and what will happen in them. Don't hesitate to repeat and repeat until you're satisfied. After all, it's *your* lab.

Third, each experiment has its own drawing to guide you in setting up the experiment. If you have some trouble arranging the equipment or following the directions, use the illustration to help you see what is happening in the experiment. The caption under the illustration will also guide you.

Fourth, we explain in a straightforward yet simple way what happens in each experiment. Our explanation is in the form of a general statement, which you can then apply to other experiments and events in your everyday life. Obviously, this statement doesn't go too deep every time. If it did, this book would be a difficult, "telling" textbook rather than an "experimenting and discovering" lab book.

However, *do not* read the explanation right away. After you have completed the "doing" part, pause for a while and consider what you have done. Just what is it that happened? What was the sequence of events? Why did it happen? In what ways is this experiment like others in this book? In what ways is this experiment related to other events and things in your everyday life?

Perhaps you won't be able to answer these questions as well as you'd like to. Nevertheless, ask them of yourself, begin to answer them, and talk them over with other people. No doubt you will find that you do have some of the answers while at the same time having some further questions. Great! It might be helpful for you to repeat the "doing" part. The talking, questioning, and repeating are all a central part of doing and learning science.

Fifth, we've extended most of the experiments by relating them to everyday events and objects—showing how the general principle used in each experiment appears elsewhere in our lives. For example, in the experiment relating the effect of air pressure on water, we point out that you apply this principle involving greater air pressure as you sip soda through a straw. The point is that science does not exist only in a laboratory. Science is alive and well and functioning in our daily lives all the time. What is more, when you develop a greater sensitivity to scientific principles as a result of using this book, we're sure that you'll be able to identify many more "scientific" things you do every day.

The purpose behind these key features of the experiments is to transform you from a spectator of science into an active discoverer. The joy of learning science comes from *observing* what you do, *thinking* about it as you answer important questions, and *figuring out why* the results happen as they do. This means that not only will you be gaining scientific knowledge and ideas, but also that you'll be learning the skills of a practicing scientist.

The experiments are so designed that you will practice observing events with *all your senses,* not just your sight. You'll practice *inferring* from what your senses tell you; that is, you'll go beyond the facts at hand to other facts and ideas. You'll practice offering possible explanations for the results of the experiment. You'll practice searching for patterns, and you'll practice stating *generalizations,* statements that connect many events and objects to the same idea. These are essential skills that every scientist learns and uses regularly.

The outcome of this experimenting and discovering approach to science is that you'll learn scientific facts, principles, and skills. At first it may seem that this outcome is slow in arriving. However, in the long run this approach offers the soundest and actually the fastest way to learn science, since you learn best what you discover by yourself. The fruit this approach bears will keep your mind alive for a long time.

If you are a young girl or boy, first skim through this book to find some experiments that catch your attention; then start wherever you want. Do the experiments and talk about them with your friends, parents, other members of your family, teachers, and group leaders. Next read the explanations and their connections with everyday life. Then try some related experiments in the same group to apply the ideas you've been thinking about and learning. Don't be afraid to try an experiment over and over again so you can better observe what's going on. The more you experiment as a scientist, the more you learn and develop your skills. Have fun and enjoy your personal laboratory as you learn science by doing science.

If you are a parent, teacher, or group leader (scout leader, for example), use this book as a guide to involve your youngsters in active experimentation. Choose experiments that you feel will appeal to your developing scientists. Let them perform the experiments themselves. Your task is to serve as guide, not "doer" or demonstrator. Youngsters need concrete experiences as a basis for understanding abstract ideas. They gain such experiences by performing the experiments themselves and using all their senses to build a foundation to support abstract ideas.

When the doing part of the experiment is over, be sure to talk with the youngsters. Start talking about what happened so you can help them develop the skill of observing. It is essential that you spend sufficient time going over what happened before going on to the explanation. If necessary ask the youngsters to repeat the experiment.

When you feel that the facts of the experiment are clear and known, then begin to talk in terms of explanations. Here you can use the connection with similar everyday events as a springboard. Remember again that it is your primary task to *guide* the youngsters. Let the experimenters offer possible explanations, for experimenters in the lab must learn to cope with their results themselves. Discuss these possible explanations and offer help when needed.

The significant element here is your guidance of the youngsters to learn science and have fun by active participation. They must do the discovering. They must be physically active in performing the experiments and mentally active in observing, inferring, explaining, generalizing, and relating. With such an approach you can use these experiments to augment other activities you are involved in with the youngsters. You can tie together those other activities by using these experiments as bridges and applications. For example, you can relate the experiments on buoyancy to an ecology activity in which you observe floating or sunken logs, pointing out that the wood floats when it is relatively dry and contains entrapped air but sinks when it becomes waterlogged. It is through this approach that you can engage the youngsters in acquiring knowledge and skills so important to modern life.

If you have comments or new experiments to suggest, we urge you to send them to us in care of Professor Ronald Hyman, Rutgers University, 10 Seminary Place, New Brunswick, New Jersey 08903. We look forward to hearing from you.

**experiments
with everyday objects**

air and
water pressure

Moving Two Apples Together without Touching Them

Equipment needed:
- two apples with stems ▪ two 1-foot lengths of string
- a support from which to hang the apples

Hang the apples about 1 inch apart. Blow hard between them and watch them move toward each other. Why do they move?

The pressure in a fast-moving airstream is less than in the still air around it. The moving air between the apples presses against them less than the still air on the opposite sides. The still air pushes the apples together.

An airplane wing works the same way. Air is forced to move faster over the top of the wing and more slowly underneath because of the shape of the wing. The difference between the higher pressure on the bottom and the lower pressure on the top of the wing holds up the plane's weight.

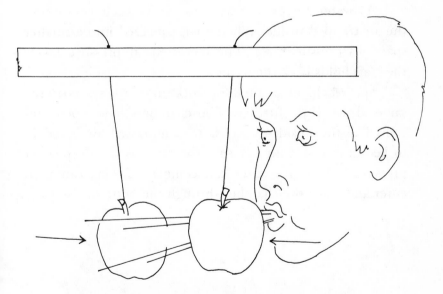

When you blow hard between the two hanging apples, they move in toward each other.

Pushing Paper Underwater
without Getting It Wet

Equipment needed:
- sheet of paper   ■ tall glass   ■ bowl of water

Crumple the paper and wedge it tightly into the bottom of the glass so that it does not fall out when the glass is turned upside down.

Now push the upside-down glass straight into the bowl of water, taking care not to tilt the glass. The water will rise a short distance into the glass, but not far enough to wet the paper. Why?

As water forces its way into the glass, it compresses the air trapped inside. When air is squeezed into a smaller space, its pressure rises. The increased air pressure inside the glass holds back the water.

Special chambers used for work under the sea carry the same idea a little further. The chambers have open bottoms like the upside-down glass. Compressed air is fed in to keep the inside air pressure the same as the deep-water pressure so that no water can come in. Divers can then enter and leave the chamber through the bottom.

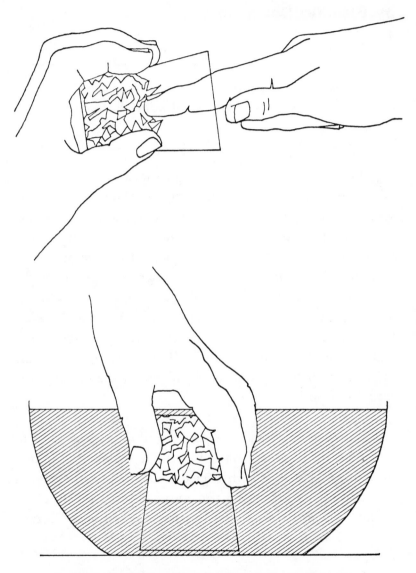

Paper crumpled and wedged into the bottom of a glass stays dry when the glass is pushed under water upside-down.

Keeping a Card from Falling
by Blowing Down on It

Push the pin halfway through the center of the card. With one hand, hold the card flat against the bottom of the spool with the pin fitted loosely inside the hole in the spool.

Now put your lips against the top of the spool and blow down hard and steadily through the hole. You will find that if you blow just hard enough, not too hard, the card will not fall when you take your hand away. Why?

The pressure in the fast-moving stream of air between the card and the spool is lower than the pressure in the still air underneath the card. The higher-pressure air underneath holds up the weight of the card and keeps the card from falling.

A perfume bottle atomizer works the same way. A fast stream of air from the squeeze bulb flows past the end of a tube whose other end is under the surface of the perfume in the bottle. The higher outside air pressure pushes perfume up into the tube and into the lower-pressure airstream. The airstream sprays the perfume out of the nozzle.

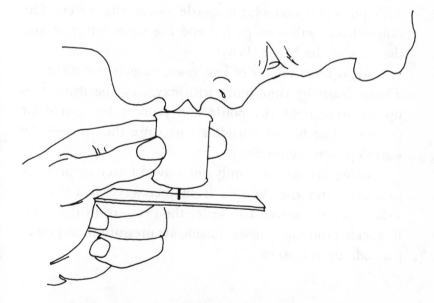

When you blow hard and steadily through the hole at the top of the spool, the card does not fall.

Making a Candle
Lift Water

Equipment needed:
- candle ■ bottle (at least 2 inches taller than the candle, with a neck wider than the candle) ■ matches
- saucer ■ water

Light the candle. Drip some melted wax into the center of the dry saucer. Quickly stand the candle straight up in the hot wax and hold it until the wax hardens. Fill the saucer half full with water.

Now place the bottle upside down over the candle with the open end of the bottle under the water. The candle flame will soon go out and the water will rise into the neck of the bottle. Why?

Air consists mostly of two gases, oxygen and nitrogen. Things burn by combining with oxygen. The flame uses up the oxygen in the bottle and reduces the inside air pressure. The higher outside air pressure then pushes the water up into the bottle.

Using up oxygen is only one way of lowering pressure to make water rise. You do it another way when you sip soda through a straw. You lower the pressure in the straw by sucking out air. Higher outside air pressure then pushes the soda up the straw.

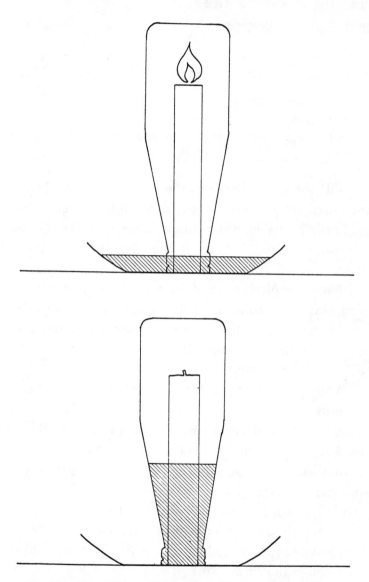

By the time the candle flame inside the upside-down bottle goes out, the water will have risen from the saucer part way into the bottle.

Making a Bottle Rise
and Fall in Water

Equipment needed:
- tall glass
- small empty bottle (from perfume or medicine)   ▪ water

Fill the glass almost to the top with water. Push the uncovered small bottle upside down into the glass, tilting the bottle to let in just enough water to make the bottle float upside down. Then fill the glass to the brim with water.

Now completely cover the top of the glass with your hand and press your palm down against the water, sealing the rim of the glass with your hand so no water spills out. The bottle will sink. If you remove your hand, the bottle will rise. Why?

Your hand pressure forces water into the bottle and the water squeezes the air into a smaller space. When enough water has been pushed into the bottle, the combined weight of the bottle, air, and water causes the bottle to sink. When you remove your hand pressure, the squeezed air expands and pushes some water out of the bottle. The bottle becomes lighter and floats.

Submarines operate the same way. Seawater is pumped in and out of special tanks to make the submarine heavier or lighter so that it can sink or rise.

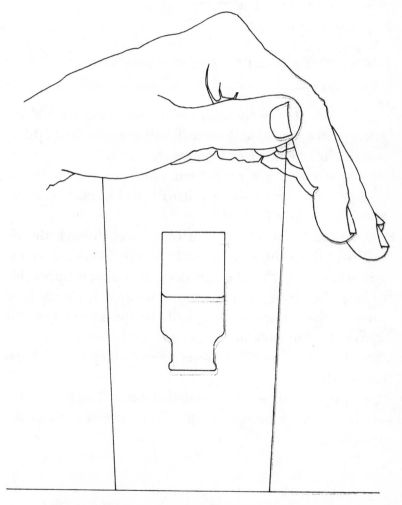

Pressing your hand against the water in the glass makes the bottle sink. Removing your hand makes the bottle rise.

Dropping a Playing Card
onto a Table

Equipment needed:
- playing cards
- small table (about the size of a chair seat)

Drop a playing card, thin edge downward, from about 3 feet above the table. You will find that the card falls to one side or the other and misses the table.

Now repeat the experiment, but this time hold the card perfectly flat. This time it will tend to float down on to the table. Why?

In the first case, the thin edge slices through the air and is easily pushed off course by slight differences in the air resistance met by the two sides. It is almost impossible to drop the card exactly straight down so that the air pressures on both sides are balanced. In the second case, air strikes the full bottom face of the card, but not the top face. This time the fall is slower, more balanced, and therefore straighter.

A parachute is like a rounded card. The broad area increases the air resistance and slows down the fall of the parachutist.

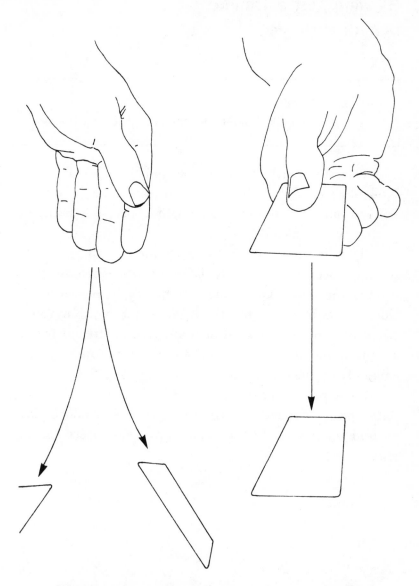

A card glides swiftly to one side or the other if dropped edge down, but falls more slowly and straighter if dropped flat.

Blowing Out a Candle
behind a Bottle

Equipment needed:
- candle ▪ candle holder ▪ matches ▪ round bottle

Place the bottle between you and the lighted candle. Now blow against the bottle toward the candle. The flame will go out even though you might expect the bottle to protect the candle. Why?

The bottle forces the moving air to split into two streams. Because the bottle is round, the streams curve around the bottle and come together again on the other side, where they blow out the flame. If the bottle were flat like a pane of glass instead of round, the airstreams would curl around the edges, but would not flow together smoothly on the other side.

Cars, planes, boats, and other vehicles are designed with special curved surfaces to allow the wind to flow around them smoothly so their movement meets less resistance.

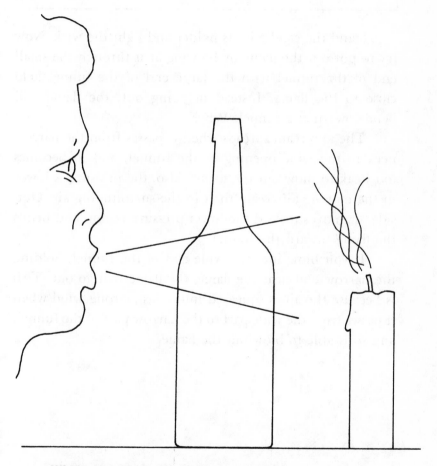

When you blow, the candle flame goes out even though it is hidden behind the bottle.

Blowing Out a Candle
through a Funnel

Equipment needed:
- candle - candle holder - matches - small funnel

Stand the candle in its holder and light the wick. Now try to put out the flame by blowing at it through the small end of the funnel with the large end of the funnel held close to the flame. Instead of going out, the flame will bend toward the funnel. Why?

The airstream spreads when it passes from the narrow neck to the wide opening of the funnel, and so becomes too weak to blow out the flame. Also, the pressure is lower in the moving airstream than in the surrounding air. Outside air moves toward the lower-pressure region and bends the flame toward the funnel.

If you blow into the wide end of the funnel, holding the narrow end near the flame, the flame will go out. This is because the air is squeezed into a fast, strong wind when it passes from the wide part to the narrow part of the funnel, and so is able to blow out the flame.

**The candle flame is almost impossible to extinguish
by blowing through the narrow end of the funnel.**

Making a Balloon
Almost Impossible to Blow Up

Equipment needed:
- balloon ■ large soft-drink bottle

Push the balloon into the bottle and stretch the open end of the balloon back over the mouth of the bottle.

Now blow hard into the balloon. The balloon will only inflate to a certain size and then stop no matter how hard you blow. Why?

As the balloon starts to fill with your breath, it takes up more space in the bottle. The air trapped between the balloon and the bottle is squeezed into a smaller space, so it rises in pressure and pushes back against the balloon. When the back pressure of the air is as strong as the pressure you can exert with your breath, you cannot blow the balloon up any further.

The same thing happens if you plug the open end of the pressure hose on a bicycle pump and try to push down the plunger. The plunger will move down until the compressed air at the bottom of the pump pushes upward as hard as you are able to push downward. You will not be able to push it down further.

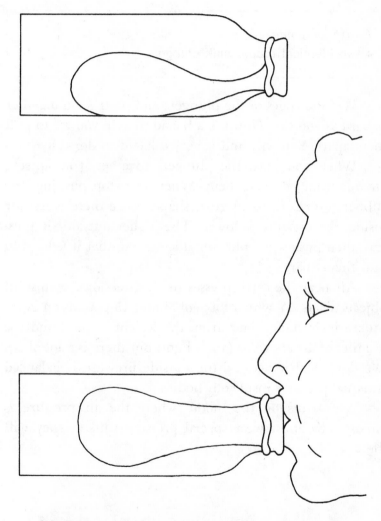

A balloon sealed into a bottle inflates only partially and then stops growing larger no matter how hard you blow.

Demonstrating
Air Pressure

Equipment needed:
- two identical rubber sink plungers

Wet the edges of the plungers and push them together as hard as you can. Now ask a friend to help you try to pull them apart. You will find it very difficult to do. Why?

When you push the plungers together, you squeeze air out from between them. When you stop pushing, the rubber returns to its original shape. Since there is less air inside, the pressure is lower. The higher outside air pressure then pushes the plungers together so that it is hard to pull them apart.

Air near the earth presses in all directions, against all objects (even us), with a force of almost 15 pounds on every square inch. This comes from the weight of the hundreds of miles of air above us (air is light, but there is a lot of it). We don't feel it because the outside pressure is balanced by equal pressure inside our bodies.

Astronauts on the moon, where the air pressure is almost zero, must wear special pressure suits or they will die.

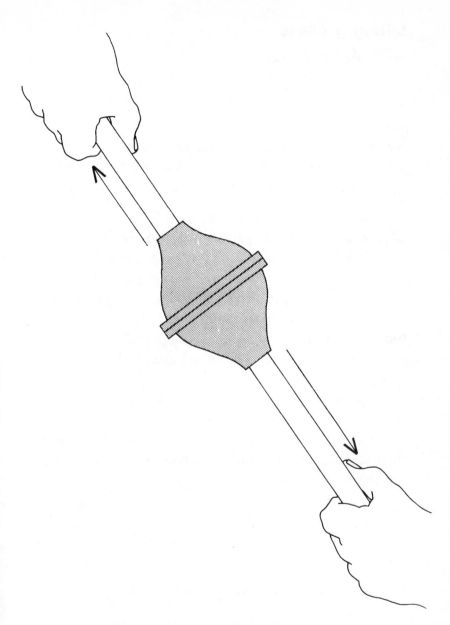

The plungers are very hard to pull apart after the air between them has been squeezed out.

Lifting a Glass
with Air Pressure

Equipment needed:
- two identical glasses ▪ small sheet of blotting paper
- water ▪ short candle with flat bottom ▪ matches

Put the candle in the bottom of one glass. Sprinkle water over the blotting paper until the paper is damp.

Now light the candle, swiftly cover the glass with the wet blotting paper, and put the second glass upside down on the blotting paper directly over the top of the first glass.

The candle flame will soon go out. If you then lift the top glass carefully straight up, you will find that the glasses are stuck together and the bottom glass will also be lifted. Why?

The burning candle goes out when it uses up the oxygen in both glasses (the damp blotting paper allows air to seep through). The air pressure inside the glasses is then lowered and the higher outside air pressure pushes the glasses together.

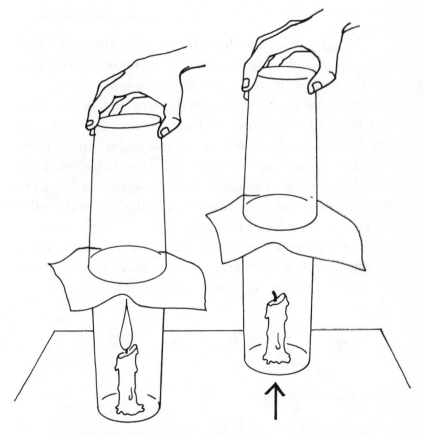

When the candle flame goes out in the bottom
glass, separated from the top glass by wet blotting
paper, you can lift the top glass and the bottom
glass will come with it.

Demonstrating that
Air Has Weight

Equipment needed:
- old, thin 1-foot wooden ruler   ▪ newspaper   ▪ table

Rest the ruler on the table so that about 4 inches sticks out over the edge. Place several sheets of newspaper over the part of the ruler resting on the table.

Now sharply hit down on the free end of the ruler with your fist. The ruler will break instead of flying up into the air with the newspaper, as you might expect. Why?

Even though air is light, it resists being suddenly pushed up by the newspaper. Therefore, the air holds the newspaper and the ruler down so that the end of the ruler breaks over the edge of the table.

The newspaper acts like the sail on a boat. Wind pressing against a sail makes the boat move in spite of the fact that the moving air is thin and invisible.

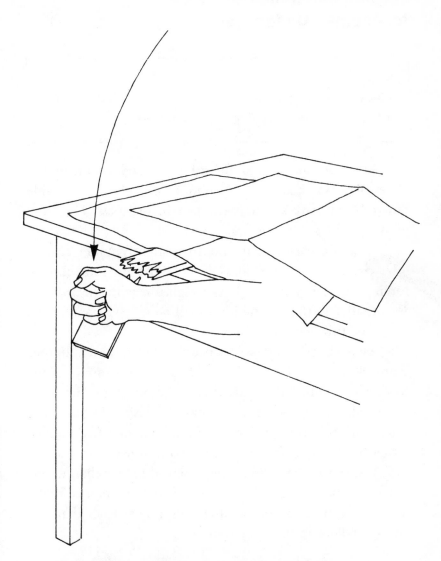

When you strike down sharply enough on the end of the ruler sticking out from under the sheets of newspaper, the ruler breaks instead of sending the newspaper flying.

Pouring Air from One Glass
to Another Underwater

Equipment needed:
- two identical glasses
- fish tank or deep sink nearly full of water

Allow one glass *(a)* to fill with water and hold it underwater. Quickly place the second empty glass *(b)* upside down in the tank so air is trapped in the glass. Hold the glass straight so that no water leaks in.

Now tilt glass *b* so its mouth is slightly under glass *a*. The escaping air from glass *b* will float up into glass *a*, forcing out the water and filling glass *a* with air instead. Eventually, all the air in glass *b* will be in glass *a*, and glass *b* will be full of water. Why?

This experiment works because air is much lighter than water and rises as bubbles. The air cannot escape from glass *b* until the glass is tilted. Then water pressure forces the air out of glass *b* and the rising bubbles are caught at the top of glass *a*. Increasing air pressure in glass *a* then pushes the water downward and out of the glass.

Just as air rises in water because it is lighter than water, warm air rises in a room because it is lighter than cold air. Therefore, the air near a ceiling is usually warmer than the air near a floor in the same room.

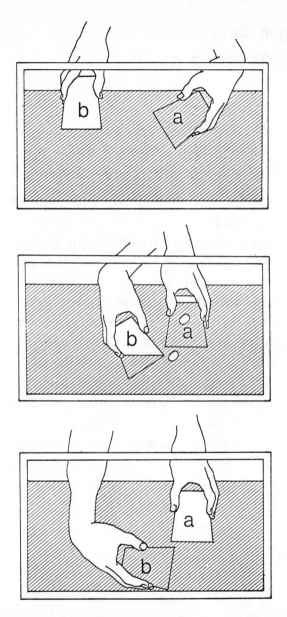

Air trapped in upside-down glass *b* escapes when the glass is tilted, rises into glass *a*, and pushes out the water.

Moving a Coin
without Touching It

Equipment needed:
- empty soft-drink bottle ■ coin ■ water

Cool the bottle in the refrigerator for at least 1 hour, then remove it and set it on a table. Quickly sprinkle drops of water around the open top of the bottle and on the coin. Place the coin, wet side down, on top of the bottle.

Now wrap your hands around the bottle, covering as much surface as possible. Soon the coin will begin to dance up and down slightly, and will keep moving for a while even when you take your hands away. Why?

The warmth of your hands expands the cold air inside the bottle. The expanding air pushes past the coin to escape from the bottle. As the escaping air lifts different parts of the coin, the coin seems to dance.

Steam expanding from boiling water in a pot makes a loose-fitting lid dance up and down like the coin in the experiment.

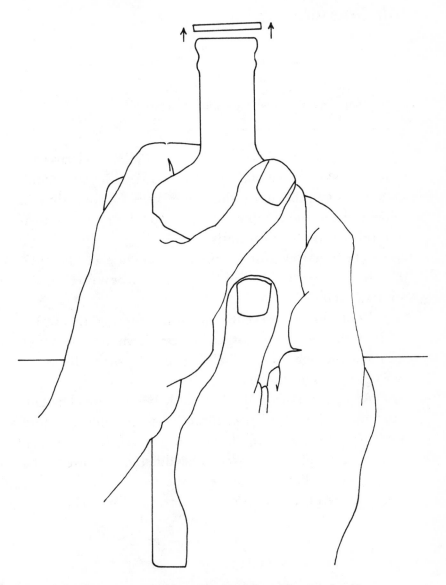

Warming the cold air in the bottle with your hands makes the coin move up and down.

Demonstrating
Air Pressure

Equipment needed:
- clear, flexible plastic tubing ▪ water

Bend the tube into a U-shape and fill it midway with water. Partially close one end of the tube with your thumb and tilt the tube until the water rises to your thumb. Now seal the end completely with your thumb and return the tube to the vertical position.

The water will stay raised in the side of the tube closed by your thumb, and the level will be lower on the opposite side. Why?

Your thumb prevents air from entering the sealed end of the tube, but air can press down on the water through the open end. The air pressure forces the water up against your thumb.

A longer U-shaped tube, called a manometer—also closed at one end and partially filled with water or liquid mercury—is actually used by scientists to measure the pressure of air and other gases. The difference between the heights of the liquid columns in the two sides of the tube is a measure of the pressure that enters the open end.

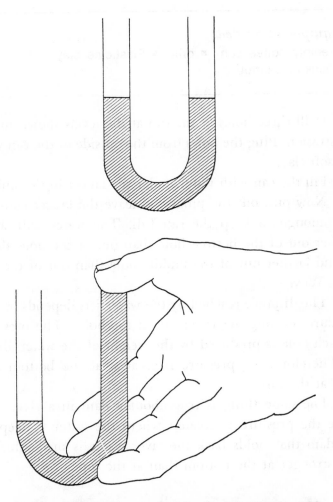

When you close one end of the U-shaped tube with your thumb after tilting the tube to fill that side with water, the water does not flow back to its original level after the tube is made vertical again.

Proving that Water Pressure Increases with Depth

> *Equipment needed:*
> - empty coffee can   ■ drill   ■ Plasticine clay
> - sink or bathtub

Drill three holes in the can at the levels shown in the illustration. Plug the holes from the outside of the can with the soft clay.

Fill the can with water under the faucet in the sink or tub. Now pull out the plugs and leave the faucet running just enough to keep the can full. The water will shoot farther out of the bottom hole than out of the hole above it, and farther out of the middle hole than out of the top hole. Why?

The distance reached by the water jets depends on the pressure pushing the water out of the holes. The pressure at each hole is produced by the weight of the water above it. Therefore, the pressure is greatest at the bottom and least at the top.

The same thing is true about a dam in a river. Because the pressure is greatest where the water is deepest, the dam that holds back the river water is made thicker and stronger at the bottom than at the top.

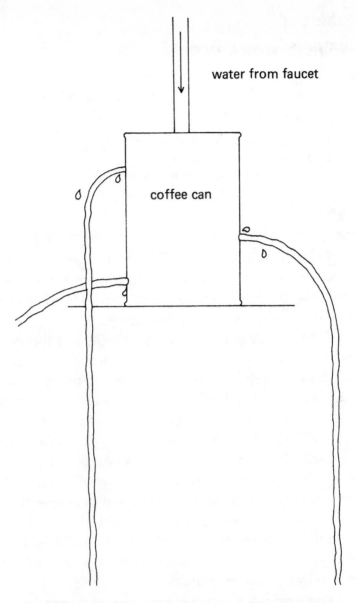

water from faucet

coffee can

Water shoots farther out of a hole near the bottom of a water-filled can than out of holes nearer the top.

Making a
Compressed-air Rocket

Equipment needed:
- narrow plastic drinking straw
- wide plastic drinking straw
- plastic squeeze bottle with a screw top
- thin cardboard ■ glue ■ scissors ■ Plasticine clay
- drill

Drill a hole in the bottle lid just wide enough to slide the narrow straw through. Leave 4 inches of straw sticking out of the lid. Seal the straw into the lid with the clay. Screw the lid tightly onto the plastic bottle.

Cut the wide plastic straw to a length of 4 inches and plug one end with the clay. Cut four identical triangles of cardboard and glue them to the sides of the wide straw at the unplugged end. The triangles should form fins that stick straight out on four opposite sides of the straw.

Now slide the wide straw over the narrow one. Give the plastic bottle a sudden sharp squeeze and the wide straw will fly off for several yards like a small rocket. Why?

When you squeeze the bottle, you compress the air inside. The air pushes hard in all directions inside the bottle, but only the large straw is free to move to let the air escape. Therefore, the compressed air pushes the large straw away and makes it fly.

The same thing happens when you pull the trigger on a BB air rifle. A spring compresses the air in the rifle. The compressed air pushes in all directions inside the rifle, but only the BB pellet can move, so it is shot out of the rifle.

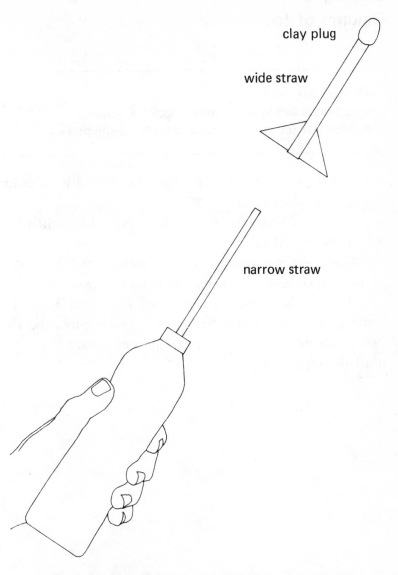

clay plug

wide straw

narrow straw

When you squeeze the plastic bottle sharply, the straw with a plugged end shoots into the air like a rocket.

Making a
Column of Ice

Equipment needed:
- soft-drink bottle with narrow neck ■ water
- deep freezer or refrigerator freezing compartment

Fill the bottle to the top with water. Stand it, with its top open, in the freezer.

When the water has frozen, a column of ice will stick out of the top. Why?

Since water expands when it freezes, the ice takes up more space than the water did before it froze.

This is the reason water pipes sometimes burst in wintertime. When water freezes in a closed pipe, the ice that forms needs more room and presses against the pipe until the pipe splits.

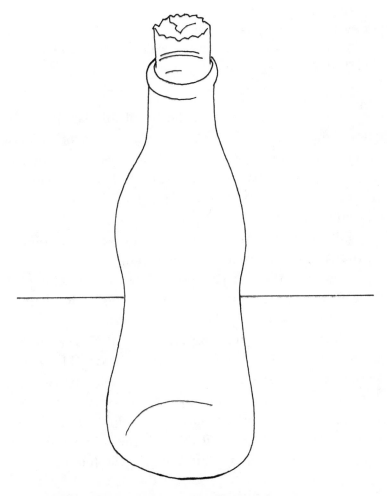

Water expands when it freezes into ice, so the ice rises from the bottle that was filled to the brim with water.

Making a Fountain
inside a Bottle

Equipment needed:
- two screw-top jars   ▪ drill
- two plastic drinking straws   ▪ water   ▪ plasticine clay
- small stand

Drill two small holes in the lid of one jar. Push one straw about 2 inches through one hole and the other straw about ½ inch through the second hole. Make an airtight seal around the straws with the clay. Fill the jar half full of water and screw the lid on tight. Cut the first straw so that only a few inches stick out of the jar.

Fill the second jar full of water. Leave the lid off and set the jar near the edge of a raised stand.

Now turn the first jar upside down and quickly push the shorter straw into the jar full of water. A fountain of water will shoot up into the upper jar. Why?

When you turn the closed jar with the straws upside down, water spills out of the longer straw. The empty space left where the water was causes the air pressure inside the jar to become lower than the outside air pressure. The higher outside air pressure pushes down on the water in the open jar and forces it through the connecting straw into the upper closed jar, thus creating the fountain.

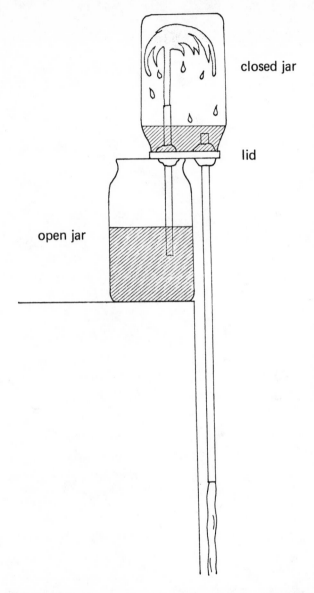

closed jar

lid

open jar

Water from the upper jar pours down out of the longer straw, and water from the lower jar rises through the shorter straw and makes a fountain.

buoyancy

Telling Bad from Fresh Eggs
without Breaking Them

Equipment needed:
- large glass container ▪ salt ▪ teaspoon
- cold water ▪ fresh and stale eggs

Dissolve one teaspoonful of salt for each half pint of cold water in the glass container. Carefully drop the eggs into the salt water.

A fresh egg (or one that has been kept refrigerated) will sink to the bottom *(c)*. An older egg that has not been refrigerated will float just below the surface *(b)*. A really bad egg will float on the surface *(a)*. Why?

Caution: Do not break a bad egg because it has a terrible smell.

The density (heaviness) of the egg becomes less as it grows old without refrigeration because it tends to dry out. The lighter it is, the more it floats in salt water.

The opposite happens when you place dry paper on water. The paper floats until it absorbs water and becomes heavier, and then it sinks.

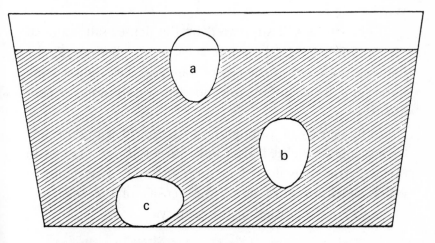

A fresh egg *(c)* sinks, a partly stale egg *(b)* half-floats, and a really bad egg *(a)* floats at the surface of salt water.

Floating an Egg in the Middle of a Glass of Water

Equipment needed:
- tall glass ▪ salt ▪ tablespoon ▪ water
- fresh egg

Put three tablespoonfuls of salt in a half-full glass of water and stir until the salt dissolves.

Now carefully place the uncooked egg in the center of the water. The egg will float because salt water is denser than the egg.

Now pour fresh water slowly on top of the egg, so that the fresh and salt water do not mix, until the glass is full. The egg will remain floating in the middle of the glass. Why?

The egg is still supported by the denser salt water at the bottom of the glass, but stays at the bottom of the lighter fresh water at the top of the glass. The lighter fresh water also floats on top of the denser salt water if not disturbed.

The density of automobile battery acid is often tested by floating a weighted glass bulb in it the way the egg in this experiment is floated in the salt water. The depth to which the bulb sinks shows the density of the acid and, therefore, the condition of the battery.

The uncooked egg floats on top of the salt water in the bottom half of the glass on the right, below the fresh water that fills the top half (the dividing line between fresh and salt water is invisible). It sinks to the bottom of the plain fresh water in the glass at the left.

Showing Why
Icebergs Float

Equipment needed:
- glass   • water   • ice cube

Put the ice cube in the glass and fill the glass to the brim with water. The ice cube will float with most of its body underwater, the way giant icebergs float in the ocean. Why?

Because water expands when it freezes, ice fills more space than the water that made it. Since ice is not as dense as water, it floats. However, because ice is only *slightly* less dense than water, only a small part of the ice cube stays above the water surface. On the other hand, most of a cork floats above the water because the cork is much less dense than water.

You can prove that ice fills less space than the water that made it by waiting until the ice cube melts. Even though the glass was completely full when some of the ice was still above the water, the water formed by the melting ice does not cause the glass to overflow.

If ice were denser than water, it would sink. Lakes would then freeze from the bottom up in the winter. There would then be no top layer of ice to skate on.

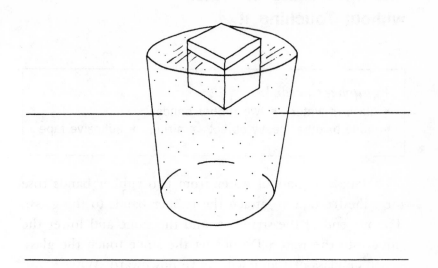

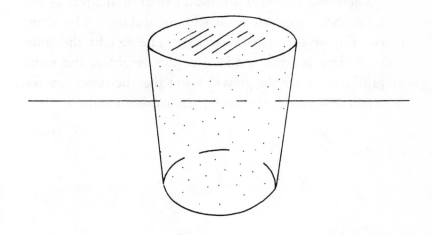

Most of an ice cube floats below the water surface. After the ice melts, the water it forms does not cause the glass to overflow, even though the glass was already full.

Moving a Glass of Water
without Touching It

Equipment needed:
- glass of water
- two rubber bands
- stone or other heavy object
- string
- adhesive tape

Hang the glass of water from two rubber bands (use the adhesive tape to attach the rubber bands to the glass). Tie one end of the string around the stone and lower the stone into the water. Do not let the stone touch the glass.

The glass of water will move downward. Why?

The stone weighs less in water than in air, just as you do (that's why you can swim without sinking). The string carries the stone's reduced weight. The weight the stone seems to lose is actually added to the weight of the water and the glass. Since the glass of water then becomes heavier, it moves downward, stretching the rubber bands.

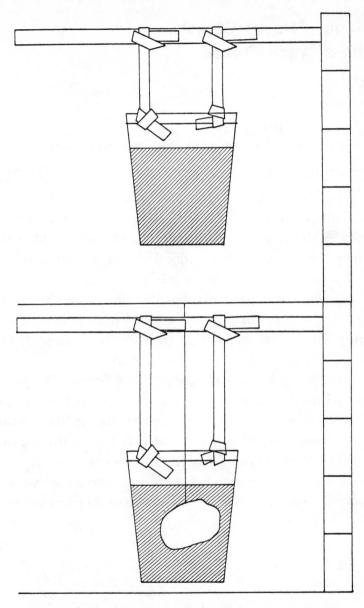

Lowering a stone into the water without touching the glass makes the glass move downward and stretch the supporting rubber bands.

Lifting a Heavy Object
with a Weak Thread

Equipment needed:
- large can of food ▪ thin thread ▪ bowl of water

Tie the thread tightly around the top rim of the can, leaving two loose ends opposite each other. Try to lift the can with the two ends of thread. The thread should break before the can leaves the ground. If the thread does not break, use a weaker thread or a heavier can.

Repeat the experiment, but this time place the can in the bowl of water before you try to lift it. Now you can gently lift the can at least an inch off the bottom of the bowl without breaking the thread. Why?

All things weigh less in water than in air. The water pushes them upward. Light things like wood are pushed all the way to the surface and float. Heavy objects like the can just become lighter, so the thread that cannot lift the can in air without breaking can lift it in water.

For the same reason, heavy air tanks carried by a scuba diver are easier to carry in the water than they are on land.

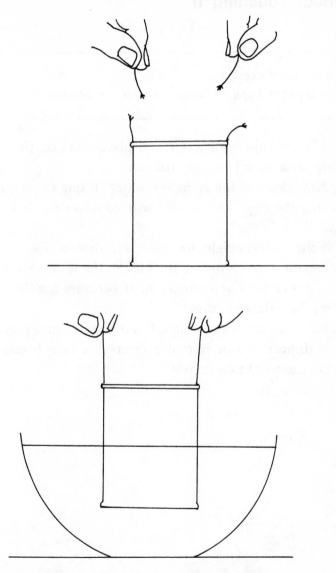

A weak thread that cannot lift a heavy can in air without breaking can lift the can in water.

Unbalancing a See-saw without Touching It

Equipment needed:
- two paper cups ▪ water ▪ ruler ▪ pencil

Fill the cups with water and place them on the ruler resting on a pencil so they balance.

Now dip one finger in the water in one cup without touching the cup. The seesaw will go down on that side. Why?

Your finger weighs less in water than it does in air. The weight it seems to lose is added to the water. Since the cup of water with your finger in it becomes heavier than before, the balance is upset.

Because everything weighs less in water, a heavy animal like a rhinoceros can rest in a river or a lake because it relieves the weight on its feet.

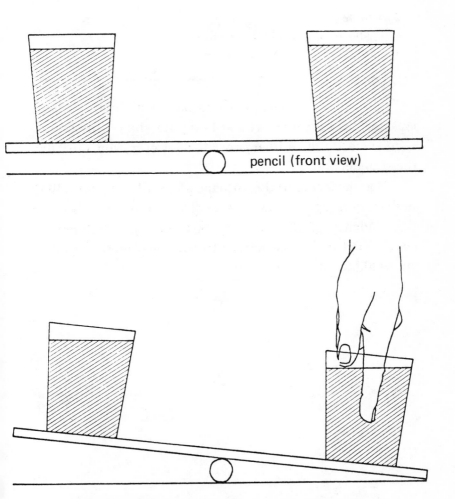

pencil (front view)

Dipping your finger into the water in one of the cups upsets the balance even though your finger does not touch the cup.

Flying an
Airplane Underwater

Equipment needed:
- plastic model airplane ▪ string
- bathtub or glass tank of water

Most plastic model airplanes cannot fly in air without crashing because they are too heavy for their wingspan to keep them up at ordinary speeds. However, you can "fly" them underwater.

Tie the string to the airplane and pull the plane along underwater. The appearance of flying is realistic because the tendency of the plane to float in water lightens its weight, and the denser water lifts the wings more than thin air would at the same speed.

When you pull a plastic model airplane underwater
with a string, it appears to fly realistically as it cannot do in air.

aking an
Underwater Color Fountain

Equipment needed:
- small bottle ▪ hot water
- red cake coloring or vegetable dye ▪ string
- large jar of cold water

Tie the string around the neck of the small bottle, fill the bottle with hot water, and add the red dye.

Now lower the bottle into the large jar of cold water. A red cloud will rise out of the small bottle like a colored fountain. Why?

Water expands when it is heated, so the red-dyed hot water is less dense than the cold water. Therefore, the hot colored water floats up as wood or cork float because they are also less dense than water.

For the same reason, the upper part of a cup of tea is hotter than the bottom part.

A colored cloud rises from a small bottle of red-dyed hot water when the bottle is lowered into a jar of cold water.

surface tension

Attracting a Matchstick
with a Lump of Sugar

Fill the bowl with water and wait until the surface becomes still and smooth. Float the matchstick near the center of the bowl. Tie one end of the thread around the lump of sugar and lower the sugar slowly into the water about 1 inch away from the matchstick.

The matchstick will soon move toward the sugar. Why?

Sugar dissolves in the nearby water. The sugar solution between the sugar lump and the match has a higher surface tension than the surrounding plain water, so it pulls the matchstick and the sugar lump together. The sugar solution is also heavier than plain water. It therefore sinks out of the way as the space between the match and the sugar becomes narrower.

If you use a piece of soap instead of a lump of sugar, a soap solution is formed. Since the soap solution has a lower surface tension than plain water, the matchstick and the soap are pulled away from each other by the plain water on both sides.

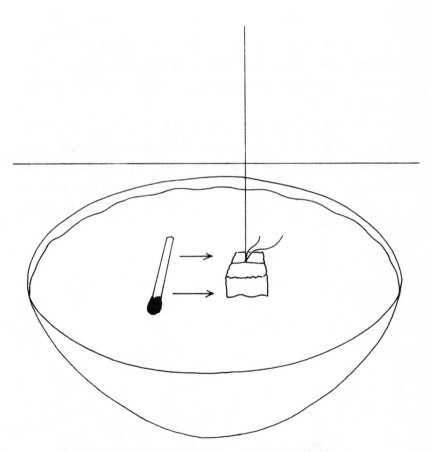

A floating matchstick moves toward a lump of
sugar placed nearby in a bowl of water.

Making a Speedboat
from a Match

Equipment needed:
- used wooden matches ■ liquid detergent ■ penknife
- bowl of water

Split the end of a match. Then, carefully place a drop of detergent in the split end and float the match on the water in the bowl. The match will move forward rapidly. Why?

The detergent lowers the surface tension of the water at the split end of the match. This detergent-water solution is pulled out of the split by the higher surface tension of the surrounding water. As the detergent-water solution moves out of the back of the match, the match recoils forward.

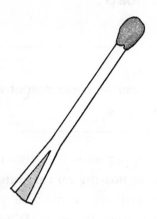

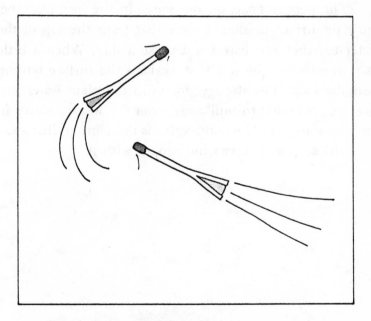

A drop of liquid detergent placed in the split end of a match makes the match move forward when it is floated on water.

Moving Pepper
with a Bar of Soap

Equipment needed:
- finely ground pepper
- clean dish of water
- wet bar of soap

Sprinkle the pepper into the dish of water until it forms a uniform layer floating on the water. Now touch an edge of the water with the soap. The pepper layer will break and run away from the soap. Why?

The pepper floats on the water in the first place because of surface tension, a force that pulls the top of the water together and makes it act like a skin. When a little soap dissolves in the water, it weakens the surface tension near the soap. The stronger forces in the plain water surface are then able to pull away from the weaker forces in the soap solution. The plain water is therefore pulled away from the soap and carries the pepper with it.

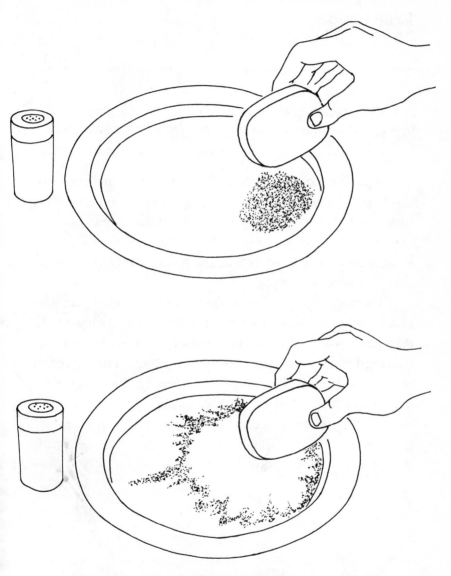

The pepper layer floating on the water breaks and runs away from the place where the water is touched by the soap.

Making Bubbles
Last Longer

Equipment needed:
- small mixing bowl ▪ water ▪ soap or liquid detergent
- glycerine (from drug store) ▪ bubble pipe or wire loop

Mix the soap or detergent and water to make a fairly thick soapy mixture. Add enough glycerine to the mixture to make it feel slightly gluey when stirred rapidly.

Now use the mixture to blow bubbles with the pipe or wire loop. The bubbles will last longer with the glycerine than without it. Why?

Forces in the water surface, called surface tension, make the water seem to have a skin. Adding soap or detergent weakens the surface tension so the skin can be stretched into a bubble without breaking. The glycerine makes the bubble last longer by keeping the water from drying quickly out of the bubble wall.

Glycerine is often used in the material you smear on your lips to keep them from drying out and becoming chapped on a cold, windy day.

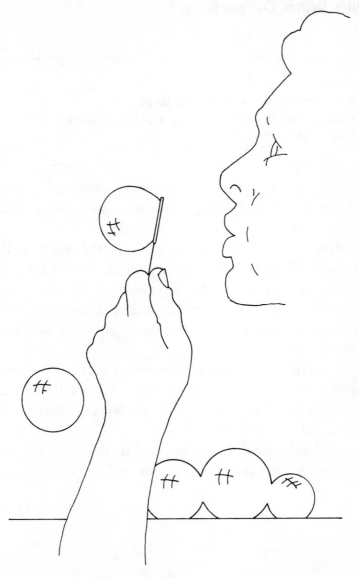

Soap bubbles last longer if you add glycerine to the soap-water mixture.

Demonstrating Water Surface Tension with Wine Glasses

Equipment needed:
- two identical, clean wine glasses
- deep bowl or sink of water   ■ penny or dime
- table or other flat surface

Hold the two glasses under water, open ends up, until full. Then, still under water, turn one upside down exactly on top of the other.

Carefully remove the glasses from the water in this position (still full of water), and place them on a flat, level surface. Now, gently tilt the top glass slightly, just enough so that you can push the coin through the narrow gap into the bottom glass. If you are careful, no water will spill. Why?

Surface tension, forces in the surface that pull the water together, makes the surface of the water act like a skin. This skin prevents the water from flowing through the narrow slit between the glasses.

Surface tension also makes bubbles possible. A bubble actually has two "skins"—one inside and one outside.

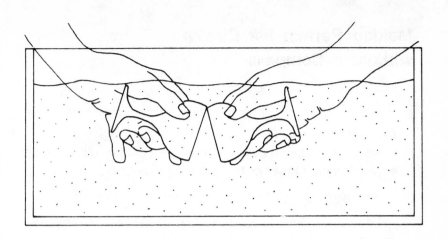

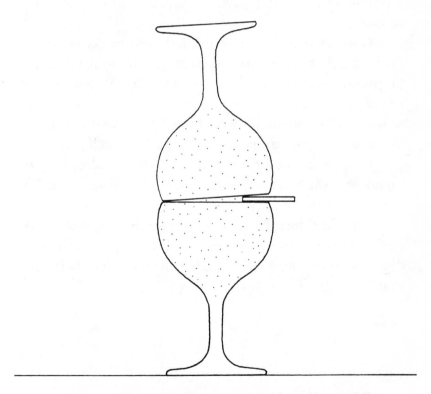

You can slip a coin between two wine glasses full of water without spilling the water.

Making Perfect Ink Circles
without a Compass

Equipment needed:
- water ■ shallow bowl or saucer ■ cooking oil
- ink ■ eye dropper or thin drinking straw

Pour some water into the bowl or saucer. Slowly and carefully spread a thin layer of cooking oil on the water surface.

Now let small drops of ink fall onto the cooking oil. (You can pick up a small amount of ink with the straw by pushing the end of the straw into the ink, sealing the top of the straw with your finger, and lifting the straw out of the ink. When you lift your finger, the ink will drop out of the straw.) The drops will form perfect circles. Why?

Because ink and oil do not mix, the ink drops try to shrink into the smallest possible space. The shape of such a space is a circle.

A similar thing happens when rain drops fall on a flat, level surface of a freshly waxed car. The wax forces the rain drops to form round balls, slightly flattened by their own weight.

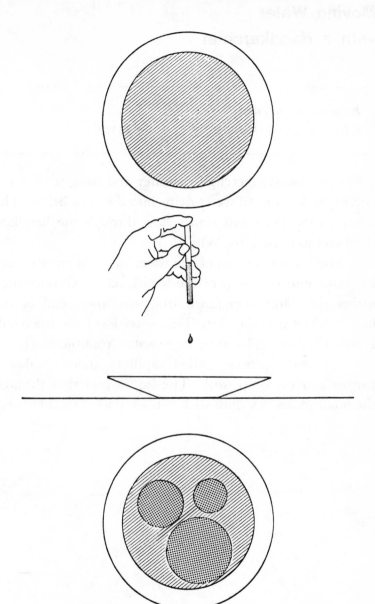

Ink drops on the surface of oil form perfect circles.

Moving Water
with a Handkerchief

Equipment needed:
- handkerchief - water - bowl - glass

Soak the handkerchief in water and hang it over the end of the bowl until water drips into the glass below. The water in the bowl will slowly flow through the handkerchief and into the glass. Why?

The surface tension of the water causes it to rise from the bowl into the cloth of the handkerchief. Gravity then makes the water drip down from the lower end of the handkerchief into the glass. This water lost from the handkerchief is then replaced by more water from the bowl.

The same process, called capillary action, makes a kerosene or oil lamp work. The liquid fuel rises through the small spaces in a cloth wick to the top where it is burned.

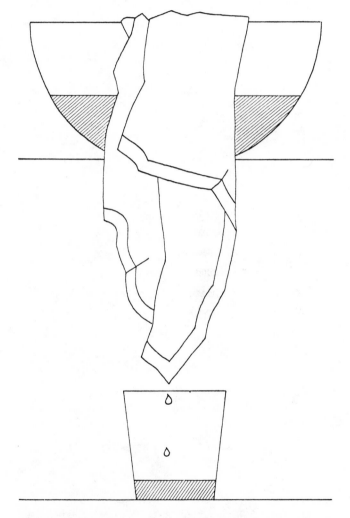

**Water slowly flows through the wet handkerchief
from the bowl to the glass.**

mechanics

Balancing a Potato
on the Edge of a Glass

Equipment needed:
- glass ▪ small raw potato ▪ two identical forks

Put the glass on a flat surface. Push the prongs of one fork upward into one side of the potato, and push the prongs of the other fork into the other side of the potato. The fork handles should stick out at the same angle on each side.

Now center the potato on the edge of the glass. Adjust the position of the potato and the forks until they balance. Why is this possible?

Balance depends not only on the weights on each side of the center, but also on their distances from the balance point. The long forks make it easier to match the weights and distances. The short potato would be almost impossible to balance by itself.

A tightrope walker in a circus often uses a long stick for balancing. The ends of the stick are like the handles of the forks, and the tightrope walker is like the potato in this experiment.

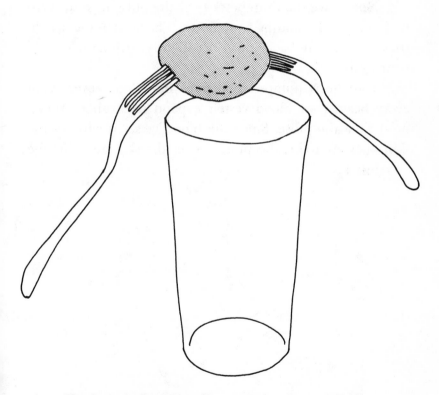

The long forks make the potato easier to balance on the edge of the glass.

Making Your Pulse Visible
with a Matchstick

Equipment needed:
- wooden matchstick ▪ thumbtack

Push the thumbtack into the bottom of the matchstick. Rest your arm on a table or other flat surface.

Now place the thumbtack over the pulse in your wrist with the match standing straight up. The head of the matchstick will vibrate back and forth slightly with each beat of your heart. Why?

Your heart pumps blood into your blood vessels with every beat. Your blood vessels expand like rubber tubes, producing the pulse. Since the blood vessel is close to the skin on your wrist, the pulse moves the skin and tilts the matchstick.

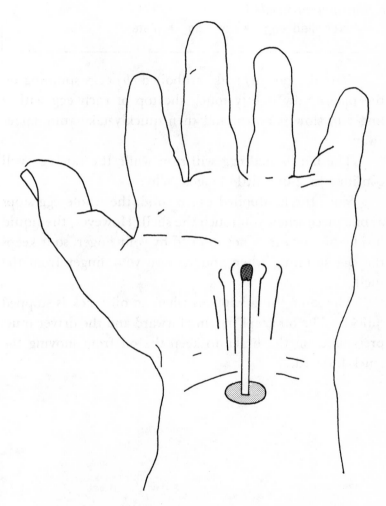

A matchstick on a thumbtack is vibrated by the pulse in your wrist.

Telling Raw from Hardboiled Eggs without Breaking the Shells

Start the raw *(a)* and hardboiled *(b)* eggs spinning on the plate. Now lightly touch the top of each egg with a finger to slow it down, and then quickly take your finger away.

The hardboiled egg will stop while the raw egg will continue to spin a little longer. Why?

Since the hardboiled egg is solid, the whole egg stops as one piece when you touch the shell. However, the liquid inside the raw egg is not stopped by your finger, so it keeps the egg spinning when you remove your finger from the shell.

The same thing happens when an oil truck is stopped quickly. The oil keeps moving forward and the driver must press hard on the brake to keep the oil from moving the truck forward.

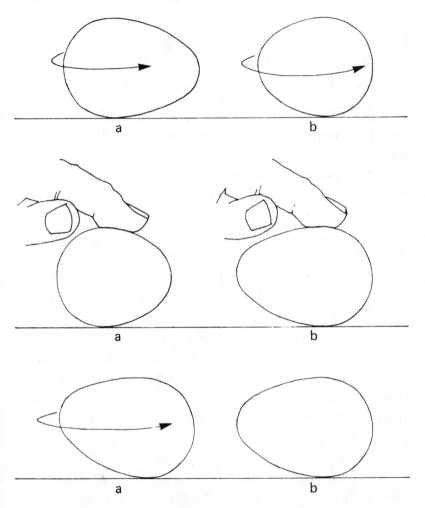

When you slow down spinning eggs with your finger, the hardboiled egg quickly comes to a stop, but the raw egg keeps spinning.

Cutting a Potato
with Paper

Equipment needed:
- straight-edged knife ■ sheet of writing paper
- raw potato ■ cutting board

Place a flat part of the potato down on the cutting board so the potato does not tend to roll. Fold the writing paper around the sharp edge of the knife.

Now press the paper-wrapped knife blade evenly and steadily down against the middle of the potato. The potato will be cut, but not the paper. Why?

The paper is forced into the potato by the knife, but the force of the sharp edge of the blade against one side of the paper is only as great as the resistance presented by the potato on the other side. Since the potato is not very hard, the knife pushes the paper through the potato without exerting enough pressure to cut the paper.

For the same reason, it is easier to push a knife through a potato than through a soft sponge. Since the potato resists, the knife can build up cutting pressure. However, the sponge just squeezes together. Therefore, the knife cannot push hard enough to cut until the sponge is squeezed as far as it can go.

You use a hard board for cutting because it gives a firm, resisting backing to the thing you want to cut.

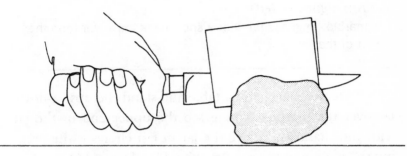

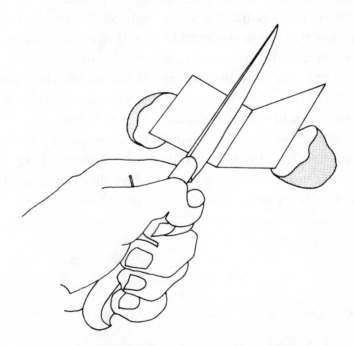

The knife pushes the paper through the potato, but does not cut the paper.

Lifting a Marble
without Touching It

Equipment needed:
- marble ▪ glass jar with a short neck narrower than the
rest of the jar

Place the marble on a table and stand the upside-down jar over the marble. Now grasp the upper end of the jar with your hand and move the jar in fast circles while pressing the open end flat against the table. The marble will be forced to roll against the inner wall of the jar. A little vibration will cause the marble to roll up above the narrower neck of the jar and into the wider part.

When you now lift the jar off the table, while continuing to rotate the jar rapidly, the marble will rise with the jar and will not immediately fall out. Why?

As the marble rolls around the jar, it presses against the glass wall. This keeps it above the shoulder between the neck and the body of the jar. The shoulder acts like a shelf to prevent the marble from falling downward.

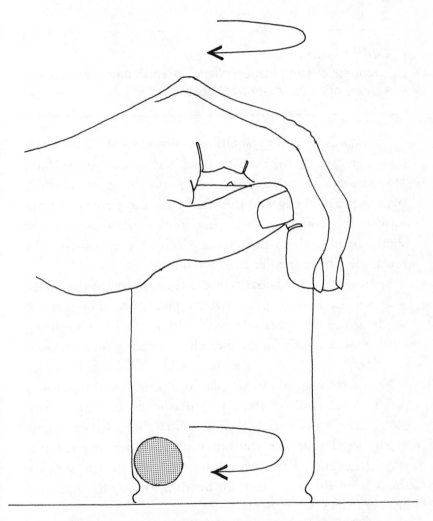

When the jar is rotated so that the marble rolls around the inside wall above the narrower neck, the marble does not fall out if the jar is lifted off the table.

Demonstrating the Strength of an Eggshell

Equipment needed:
- four equal-sized half-eggshells
- small pair of scissors
- piece of cloth
- several books

Make four half-eggshells as follows. Gently break one end of each of four eggs by tapping it against a hard surface. Do not crack the whole shell. Scoop out the eggs. Carefully trim the ragged edges of the shell with the scissors to form four half-eggshells with even, circular bottoms. Set the shells, open end down, on the cloth at the corners of a rectangle a little smaller than the books.

Now stack the books one at a time on top of the shells. You will be able to place twenty pounds or more on the shells before they break. Why should the shells be so strong when you can easily break them by squeezing them in from the sides?

When the eggshells are placed in this way, the pressure of the books follows the curve of the dome shape down along the wall of the shell. The dome also helps each part of the shell reinforce the other parts. When pressed this way, the eggshells are very strong. However, if you squeeze them from the sides, you are bending the shells, and they are very weak when bent.

Architects often use dome shapes in the design of large buildings because domes can carry more weight than a flat roof of the same size.

Half-eggshells carry a heavy load of books before breaking.

Breaking Either One
of Two Strings

Equipment needed:
- thin string ▪ heavy string ▪ heavy book

Tie the thick string around the book. Then tie one thin string at the top and the other thin string at the bottom. Hold the book up by the top string.

If you now pull down steadily and hard enough on the lower string, the top string will break. Now repeat the experiment with new strings and jerk the lower string down sharply. The lower string will break instead of the top string. Why?

In the first case, the top string carries the weight of the book in addition to the force of your pull. The lower string is stretched only by your pull. Since the top string has the heavier load, it is the one that breaks.

In the second case, the heavy book cannot move as fast as you jerk on the string. Therefore, it does not pass all the strength of your sudden pull on to the upper string. Since almost all the force of your hard pull is taken up by the lower string, it is the lower string that breaks.

This is why ropes used to lift loads on a cargo ship and cables used to raise and lower elevators in a tall building must be extrastrong. They must carry not only heavy weights, but must be able to withstand sudden hard pulls.

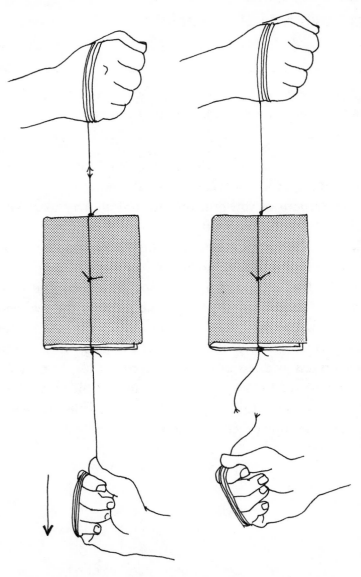

When you pull down steadily, the upper string breaks. When you jerk the lower string down suddenly, the lower string breaks.

Making a Cord
Impossible to Pull Straight

Equipment needed:
- heavy book ▪ 6-foot length of strong cord

Tie the middle of the cord around the book. You hold one end while a friend holds the other end.

Now both of you pull on the cord and try to make it perfectly straight. No matter how hard you pull, the book will always make the cord sag slightly in the middle. Why?

The pull on each end of the cord can be considered as really divided into two pulls, one horizontal, the other vertical. The horizontal parts of the pulls on the two ends are in opposite directions and cancel each other. The vertical parts lift the book. The vertical part of each pull becomes less and less as the rope becomes more and more horizontal. Therefore, it becomes impossibly difficult to pull hard enough to lift the book the last little bit needed to make the rope exactly straight.

No matter how hard you pull on the ends of the cord, the book makes it sag in the middle so the cord cannot be pulled straight.

Lifting an Apple
without Touching It

Equipment needed:
- 3-foot length of thick string
- small spool
- large spool
- apple

Tie one end of the string to the small spool. Pass the other end through the large spool and tie it to the apple.

Now hold the large spool in one hand and start to swing the small spool around in a circle. Clamp the string in the large spool with your thumb until the small spool picks up speed. Then move your thumb away to free the string.

As the small spool goes faster and faster, the string will slide up through the large spool and lift the apple. Why?

The small spool tries to swing in a wider circle as it moves faster, pulling on the string in order to do so. When the speed is fast enough, the pull becomes strong enough to lift the apple at the other end of the string.

You feel the same kind of force that lifts the apple when you ride in a car as it goes around a curve. You are thrown toward the outside of the curve.

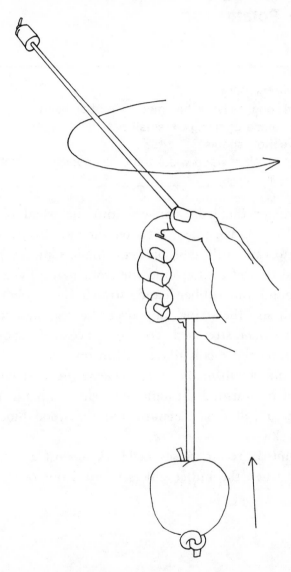

When you swing the small spool around in a circle fast enough, the string slides up through the large spool in your hand and lifts the apple.

Demonstrating Recoil Action
with a Potato

Equipment needed:
- three long nails ▪ hammer ▪ rubber band
- short piece of string ▪ small potato
- two cotton spools
- small block of wood wider than the spools ▪ matches

Hammer the nails partway into the wood block to form a "V." Leave space between the two front nails so the potato can easily pass between them. Cut the rubber band and tie the ends around the front nails. Tie a string loop around the rubber band, stretch the rubber band back, and slip the string loop over the rear nail to keep the rubber band stretched. You now have a slingshot. Rest the block on the spools like a wagon on wheels.

Set the potato in the sling. Release the stretched rubber band by burning through the string with the match. The potato will shoot forward and the wood block will recoil backward.

A similar recoil causes kickback against a rifleman's shoulder when the bullet shoots forward out of the rifle.

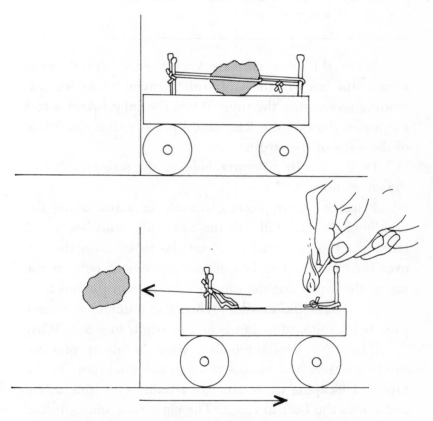

Shooting a potato forward with a slingshot mounted on a wood block riding on spools causes the block to recoil backward.

Making a
Simple Steamboat

Equipment needed:
- aluminum cigar tube with screw-on cap   • small drill
- empty sardine can   • two pipe cleaners   • hot water
- three short pieces of candle   • matches
- bowl of water

Wrap the pipe cleaners very tightly several times around the ends of the cigar tube, leaving short legs extending down from the tube. When the tube is laid across the top of the can, the legs should press against the inside of the walls of the sardine can.

Drill a small, off-center hole in the screw-on cap of the cigar tube.

Place the short pieces of candle in a line along the middle of the can. Fill half the cigar tube with hot water, screw the top on tightly, and set the tube across the can over the candles. The hole in the cap should be near the top of the cap. Float the can on the water in the bowl.

Now light the candles. When the water in the cigar tube boils, the sardine-can boat will begin to move. Why?

The steam from the boiling water builds up pressure inside the tube and escapes through the small hole in the cap. The escaping steam acts like a bullet shot from a rifle and causes the boat to recoil. The cigar tube moves in the opposite direction, carrying the boat with it, just as the rifle butt recoils against the rifleman's shoulder.

The same kind of recoil is used to send moon rockets away from the earth. Hot gases from burning fuel rush out of the rocket's tail, and the recoil drives the rocket upward.

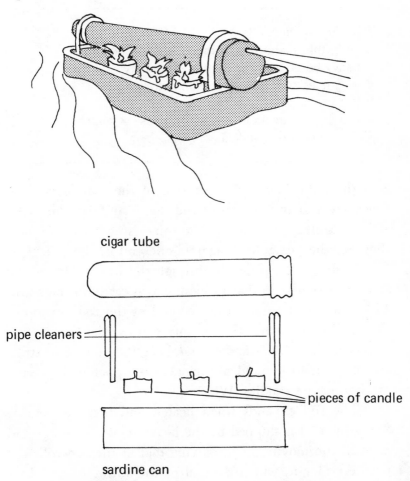

cigar tube

pipe cleaners

pieces of candle

sardine can

When steam from boiling water inside the cigar tube escapes through the small hole in the cap, the sardine-can boat moves in the opposite direction.

Proving that Metal Expands When Heated

> *Equipment needed:*
> - long, thin metal rod or metal knitting needle
> - heavy book, such as a large dictionary
> - long needle or pin ▪ drinking straw
> - two equal piles of books
> - two small mirrors or pieces of glass ▪ matches
> - three or more candles

Stack the books on each side of the row of candles. Place the mirrors on top of the books and lay the metal rod (or knitting needle) on the mirrors and over the candles. Set the heavy book against one end of the rod.

Push the sharp needle through the straw. The needle should fit tightly so that it doesn't turn easily in the straw. Place the needle under the free end of the rod, on top of the mirror, with the straw pointing upward.

Now light the candles and heat the rod. The straw will turn like the hand of a clock, the upper part moving away from the candles. Why?

The rod expands when heated and becomes longer. Since one end is stopped by the heavy book, the other end does all the moving. The moving end of the rod rolls the needle under it, which then turns the straw.

The expansion of metal when heated (and contraction when cooled) is often used in dial-type thermometers. When the temperature rises, a thin metal spring expands and turns a pointer to the number of the dial that indicates the temperature. When the temperature falls, the metal shrinks and moves the pointer in the opposite direction.

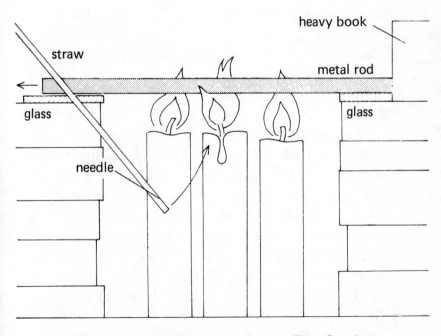

When the rod is heated by the candles, the straw turns.

Seeing
Sound Vibrations

Equipment needed:
- empty tin can ▪ cardboard ▪ lamp or flashlight
- can opener ▪ balloon ▪ glue
- ¼-inch piece of mirror ▪ rubber bands

Remove both ends from a clean, empty tin can. Cut the neck off a balloon and stretch the remaining bottom tightly over one end of the can. Hold it in place with rubber bands. Glue the small mirror onto the stretched balloon about a third of the way in from the edge of the can.

Now shine the light from the lamp or flashlight on the mirror at an angle, and place the cardboard so that it catches the reflection as a spot of light. If you now sing or shout into the open end of the tin, the spot of light will vibrate quickly back and forth. Why?

Sound is made up of vibrations. As sound travels through the air, the vibrations are passed from one air particle to the next. The air in the can passes the vibrations to the stretched balloon which wiggles the mirror and causes the reflected spot of light to move.

The diaphragm in a telephone receiver changes the sound of your voice into electrical signals just as the stretched balloon and mirror change the sound into moving light.

100

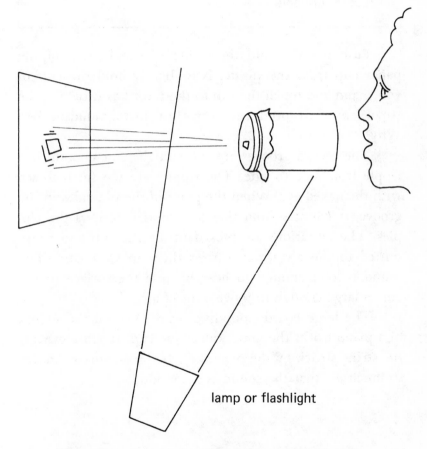

lamp or flashlight

When you sing or shout into the open end of the can, the reflected spot of light moves rapidly back and forth.

Making a Loudspeaker from a Paper Cup

Equipment needed:
- straight pin   ▪ paper cup   ▪ old phonograph record
- record turntable

Push the pin into the center of the bottom of the paper cup from the inside. Now lightly hold the cup in your hand and touch the pin to the record as it turns. The paper cup will produce music just like a loudspeaker. Why?

The record grooves are not straight and smooth, but ripple from side to side. The ripples are too small to see with the naked eye. When the point of the pin rides in the groove, it vibrates from side to side as it follows the ripples. The vibrations are picked up by the cup and transmitted to the air which carries them to your ears. The sound is loud enough to hear because the surface of the cup is large enough to move a lot of air.

The large board (sounding board) behind the strings of a piano works the same way as the cup. It vibrates with the string struck by the piano key. Its broad surface moves so much air that the sound is made loud.

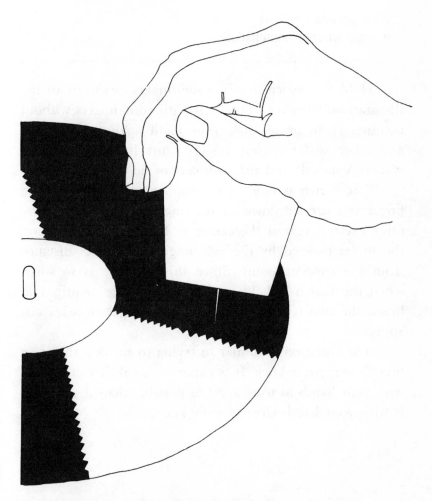

**When a pin stuck into the bottom of a paper cup is
held in the groove of a turning phonograph record,
the paper cup produces music.**

Holding a Matchstick
So It Cannot Be Broken

Equipment needed:
- used wooden matchsticks

Hold the matchstick in your fingers as shown in the illustration. Now try to break it with your fingers without bending them or pressing your hand against a table or any other surface. Your fingers must be kept close together. You will find that you cannot do it. Why?

The match is like a lever supported by your middle finger and pressed down by the fingers on each side. The force exerted against the center of the match depends on the forces exerted by the side fingers and their distance from the support point. Since this distance is so small when the match is held as shown, the force required to break the match is greater than your finger muscles can apply.

The situation is similar to trying to break a thin tree branch over your knee. It is easier to break if you hold it with your hands as wide apart as possible than if you hold it with your hands close to your knee.

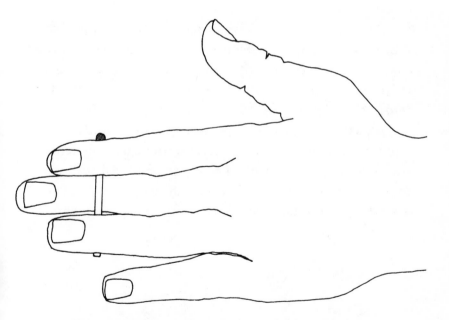

When you hold a matchstick as shown above, you cannot break it with your fingers.

Knocking Four Eggs
into Four Glasses

Equipment needed:
- four wide-mouthed glasses of water
- light, flat tray with smooth bottom
- four cardboard tubes (insides of toilet paper rolls)
- four hardboiled eggs

Place the tray, cardboard tubes, and eggs (large ends down) over the glasses of water, as shown in the top illustration. The eggs should be exactly over the centers of the glasses.

Now give the tray a quick, sharp blow with your hand, pulling your hand back before it hits the glasses. If struck exactly right, the tray will slide off to one side, dragging the tubes out from under the eggs, and the eggs will drop into the four glasses.

Practice is required to do the trick successfully. Wide-mouthed glasses are needed because the eggs will tend to turn sideways as they drop. The glasses are filled with water to give them weight so that they will not be knocked over when the tray slides over them.

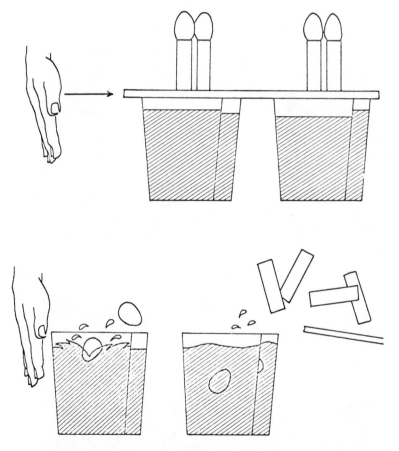

When you hit the tray sharply out from under the eggs, the eggs drop straight down into the glasses of water.

chemistry, colors, and candles

Making a
Moisture-detecting Fish

Equipment needed:
- small sheet of cellophane   - scissors

Cut a small fish shape about ¾ inch wide by 2 inches long from a thin sheet of cellophane. You must use cellophane rather than plastic food wrap. Cellophane absorbs water more easily than food wrap. Plastic food wrap is made especially less absorbent to prevent food from drying out.

Now place the cellophane fish on the palm of your hand. The fish will curl away from your skin. Why?

The side of the cellophane touching your palm absorbs perspiration (water and salts) and expands. Since the other side stays dry, it does not expand. Therefore, the fish curls upward.

Materials that expand when they become moist and shrink when they become dry are often used in scientific instruments to indicate the humidity (wetness) of the air. Human hair is one such material.

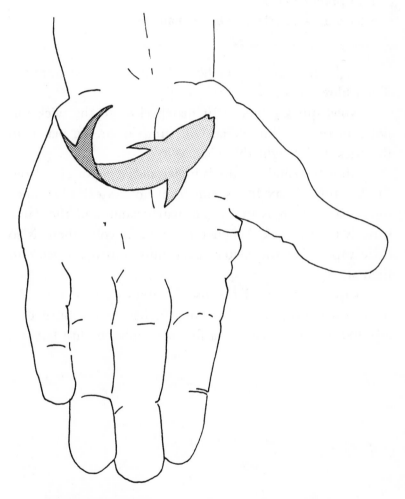

A cellophane fish curls upward when placed on your moist palm.

Lighting a Candle
without Touching It

Equipment needed:
■ candle ■ candle holder ■ matches

Light the candle and let it burn for a few minutes. Then blow it out.

Now, quickly hold a lit match close to the wick but not touching it. The flame will jump from the match to the wick and relight the wick. Why?

After the candle flame is blown out, a vapor rises from the hot liquid wax like steam from hot water. The vapor burns easily so it is lit by the match flame and the flame spreads to the wick. When the candle is cold, there is so little vapor that the match flame must touch the wick to light it.

Vapor burns well because it mixes quickly with air. That is why oil in an oil-burning furnace is not burned as a liquid, but is sprayed as a fine mist into the fire box.

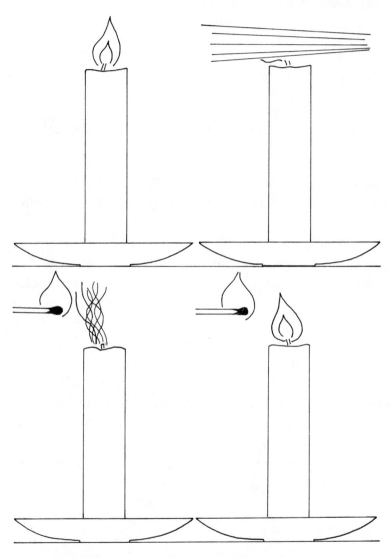

A candle that has burned for a few minutes before being blown out is relit by holding a lit match near the wick, but not touching it.

Turning a White Rose
Red and Blue

Equipment needed:
- red and blue inks or food colorings
- white rose, dahlia, or carnation
- two small containers ▪ knife or scissors ▪ water

Mix 1 part red ink (or equivalent food coloring) with 3 parts water and pour the mixture into one container. Do the same with the blue ink or food coloring and pour the second mixture into the second container.

Carefully split the stem of the white flower in half along its length and place one half of the stem in the blue mixture and the other half in the red mixture.

The two parts of the stem will soon be colored by the mixtures and, after several hours, the white flower will have become half red and half blue. Why?

The colored water rises through narrow tubes in the stem through which the flower usually takes in the water and food that keep it alive. The colors stay in the petals while the water evaporates.

The experiment also shows that different parts of the stem feed different parts of the flower.

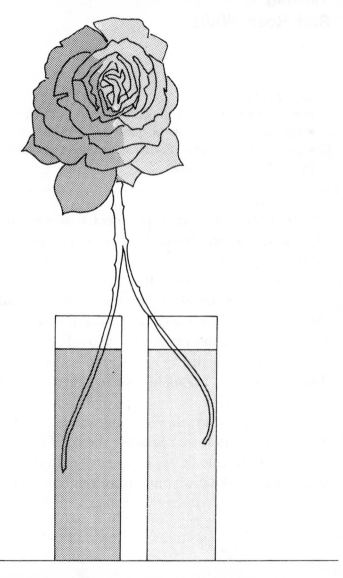

A white rose with half of its split stem in red-colored water and the other half in blue-colored water becomes half red and half blue.

Turning a
Red Rose White

Equipment needed:
- long-handled ladle with curved end ▪ teaspoon
- red rose ▪ matches ▪ glass jar with lid ▪ string
- sulfur (yellow sulfur powder is available at hobby shops that sell chemistry sets)

Important: Do this experiment outdoors because you must burn sulfur. Burning sulfur smells bad and is unhealthy if you breathe too much of it.

Tie one end of the string around the stem of the rose.

Put a teaspoonful of the sulfur into the ladle. Light the sulfur with the match and hang the ladle inside the jar, as shown in the illustration. Now lower the rose into the jar and put the lid on to keep the fumes from escaping. The rose will soon become pale and then turn white in a few minutes. Why?

Burning sulfur forms a gas called sulfur dioxide, which is a chemical that bleaches the color out of the rose.

Sulfur dioxide is an important industrial bleaching agent used to take the natural color out of things which are then sometimes dyed another color.

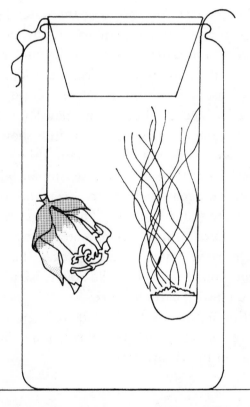

A red rose hung inside a jar with burning sulfur soon turns white.

Making Water
Wetter

Equipment needed:
- glass of water ■ finely powdered sulfur
- Calgon or similar dishwashing aid dissolved in water, or a photographic wetting agent

Sprinkle some sulfur on top of the water in the glass. Although sulfur is heavier than water, the powder stays on top because the particles remain dry and rest on the "skin" produced by the water's surface tension.

Now add a few drops of the Calgon solution (or photographic wetting agent). The sulfur powder will start to sink to the bottom of the glass like a little snowfall. Why?

Plain water does not wet certain things, including sulfur, especially if they are protected by an oily or waxy surface. Wetting agents help water overcome this protection. They make water "wetter." The wetting agent in this experiment allows a thin layer of water to coat the sulfur particles. Once the water creeps up over the top of the particles, the powder is below the water "skin" that held it up, so the powder particles sink.

A wetting agent is often added to cleaning solutions to make sure that they completely surround the dirt and carry it away in the rinse water.

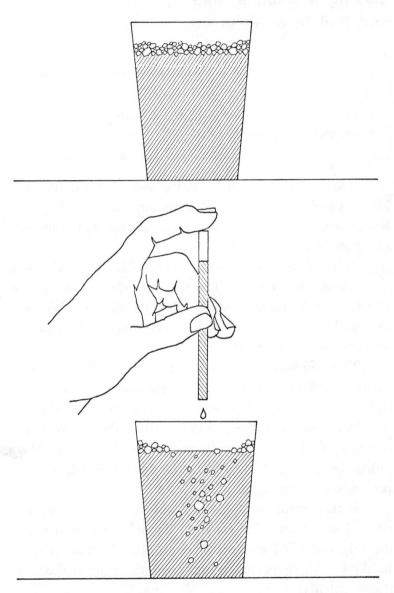

Sulfur powder stays on top of the water until a wetting agent is added.

Making Mothballs Rise and Fall in a Glass

Equipment needed:
- glass ▪ water ▪ tablespoon ▪ baking soda
- vinegar ▪ mothballs

Add a tablespoonful of baking soda and a tablespoonful of vinegar to a glass of water and stir until all the baking soda has dissolved. Put three or four mothballs into the glass. They will sink.

Bubbles will form around the mothballs and the balls will rise to the surface. They will stay there a short time and then sink. More bubbles will form around them and they will rise again, only to sink once more and repeat the up-and-down motion. Why?

The bubbles are carbon dioxide gas formed by a chemical reaction. They cling to the mothballs and their light weight makes the mothballs float to the surface of the water. At the surface, some of the bubbles break and the weight of the mothballs makes the balls sink. More bubbles then attach themselves to the mothballs and the process starts over again.

Sunken ships are sometimes raised by the same principle. Large bags, like oversized balloons, are attached to the ship and filled with air. The air-filled bags act like the bubbles in the experiment and cause the ship to float like the mothballs.

In a mixture of water, vinegar, and baking soda, mothballs rise and fall as gas bubbles form around them underwater and break when they reach the surface.

Coloring Liquids Red, Violet, or Green with the Same Dye

Cut a cabbage leaf into small pieces and soak the pieces in the cup of boiling water for 30 minutes. The water will turn violet.

Put the plain water in glass *a*, the vinegar in glass *b*, and the water–baking soda solution in glass *c*. All three colorless liquids will look the same.

Now drop some cabbage water into each of the three glasses. The water in glass *a* will turn violet like the cabbage water, but a little paler. The vinegar in glass *b* will turn red. The baking soda solution in glass *c* will turn green. Why?

Plain water only dilutes the cabbage water, but does not change its color. Vinegar is an acid and reacts chemically with the cabbage dye to form a new substance whose color is red. The baking soda is alkaline (the opposite of acid) and reacts with the cabbage dye to form a different substance whose color is green.

When two or more chemicals react to form different compounds, the resulting color is often very different from the colors of the original ingredients. A familiar example is the red rust that appears when your sled runners combine with invisible oxygen in the air.

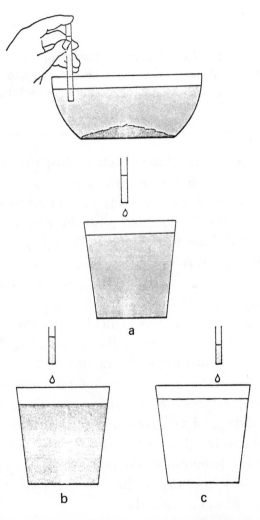

Drops of violet-colored cabbage water turn plain water violet, vinegar red, and baking soda solution green.

Making a
Fire Extinguisher

Equipment needed:
- glue ■ piece of cardboard ■ screw-top jar
- short piece of hose ■ vinegar ■ bicarbonate of soda
- spoon ■ candles ■ candle holders ■ matches
- drill

Glue a small cardboard shelf against the inside wall of the jar about 2 inches from the top.

Drill a hole in the lid of the jar just large enough to fit tightly around the hose and push the hose into the lid.

Half-fill the jar with vinegar. Do not wet the cardboard shelf. Now carefully spoon bicarbonate of soda onto the shelf, taking care not to let any powder drop into the vinegar. Screw the lid tightly on to the jar.

Light the candles. Now quickly turn the jar upside down. Foam will shoot out of the hose and will put out the flames when directed at the candles. Why?

The chemical reaction between bicarbonate of soda and vinegar produces carbon dioxide gas. The gas bubbles turn the mixture of vinegar and bicarbonate of soda into a foam and also build up pressure inside the jar. The pressure forces the foam out of the hose. Since carbon dioxide pushes air away from the candles, and the candles need air to burn, the flames are smothered.

Many fire extinguishers work the same way, but may use different chemicals.

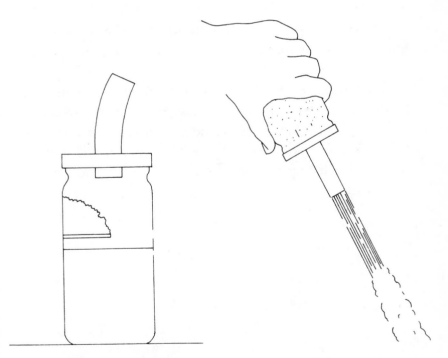

Foam from a mixture of vinegar and bicarbonate of soda smothers a candle flame.

Making a
Hollow Candle

Equipment needed:
- candle ▪ flathead screw ▪ glass of water
- matches

Screw the flathead screw into the bottom of the candle so that the candle sits upright on the bottom of the glass of water. A small part of the candle should stand above the water.

Light the candle. The candle will burn in the middle, but the wax on the outside will remain and the candle will become hollow. Why?

A candle keeps burning because more wax is melted by the heat of the flame and rises up the wick to feed more fuel to the fire. Since the outside of the candle is cooled by the water, it does not melt. Only the wax in the middle is used up during the burning, so the candle becomes a hollow tube.

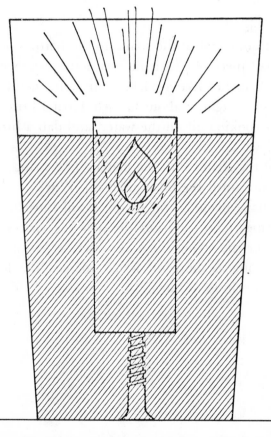

A candle becomes a hollow tube when it burns
while standing in a glass of water.

Making a Candle
without Wax

Equipment needed:
- shallow metal or glass box with a flat bottom and open top ■ short piece of string heavy enough to stand upright ■ plasticine clay ■ water ■ cooking oil ■ matches

Attach one end of the string to the bottom of the box with a small piece of plasticine clay, letting the string stick up about 2 inches like a wick. Pour water gently around the wick to a height of about ¼ inch. Float a little less than ½ inch of cooking oil on the water, and dab a little oil on top of the wick.

Light the string with the match. The string will burn like a candle until the oil is used up. Why?

You have made an oil lamp. As oil burns at the top of the string, more oil rises through the spaces between the fibers of the string the way water seeps into a paper towel. As long as more oil is left to feed the flame, the wick will continue to burn.

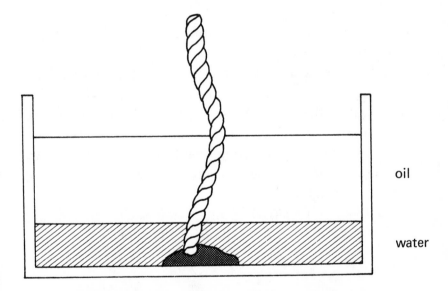

oil

water

When you stand a piece of string upright in a pool of cooking oil (the bottom water layer only protects the container) and you light the exposed end, it burns like a candle wick.

Making Your Own Watermarks on Paper

Equipment needed:
- writing paper   ▪ water
- toothpick or empty ballpoint pen
- mirror or sheet of glass

Soak the paper in water until it feels soggy. Lay the wet paper flat on the glass. Write your initials on the paper with the toothpick or empty ballpoint pen, being careful not to tear the paper, which is weak when wet.

When the paper has dried, it will seem blank, but if you hold it in front of a light you can see your own watermark. Why?

By pressing on the wet paper with the sharp point, you disturb the pattern of the fibers that make up the paper. After the paper dries, the fibers do not return to their original condition. The watermark is not easy to see with light reflected from the front of the paper. It becomes visible when light passes through the paper from behind.

Machines automatically put a watermark into high-quality paper during the making of the paper.

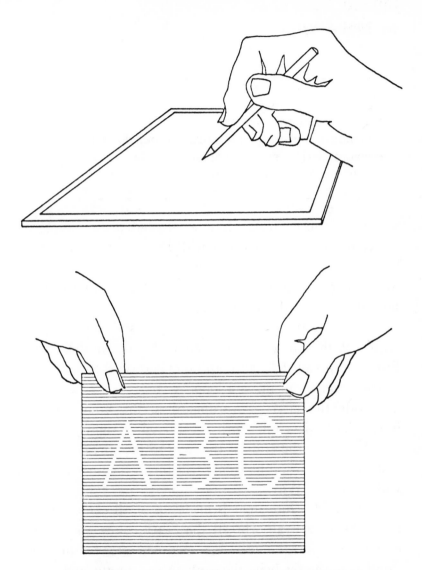

Writing with a sharp point on wet paper makes a
mark that is almost invisible when the paper dries.
The mark can be seen if the paper is held in front
of a light.

Mixing Colors
to Make White

Equipment needed:
- stiff cardboard ▪ scissors
- compass (for drawing a circle) ▪ pencil ▪ ruler
- colored paints and paintbrush
- thin, strong thread ▪ pin

With the compass, draw a 3-inch-diameter circle on the cardboard and cut it out with the scissors. Draw three lines through the center of the circle, making six equal wedge-shaped sections. Color each section with the color indicated in the illustration.

Now use the pin to make two holes in line with the center of the circle, $\frac{1}{4}$ inch on each side of the center. Push the thread through both holes and tie the ends of the thread together to make a loop.

Slide the circle of cardboard to the middle of the loop. Turn the circle until the thread is tightly twisted.

Now pull the ends of the loop so the threads untwist and spin the card. The colors will blend together and you will see white. Why?

Ordinary white light is made up of all colors. A rose is red because it absorbs all the other colors and reflects only the red part to our eyes. Sometimes, after a rain, water droplets in the sky bend the sunlight passing through them. Since light rays of different colors are bent by different amounts, the colors are spread out into a rainbow.

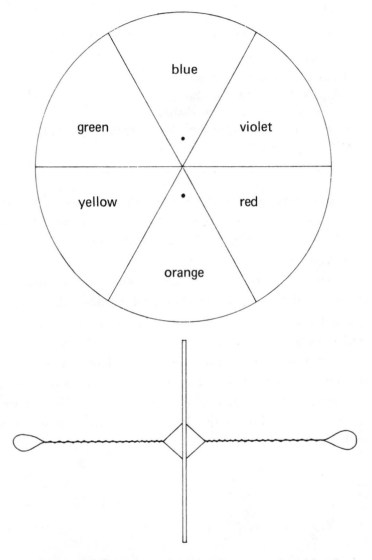

When you spin the circle, the different colors blend into each other and you see white.

Painting Pictures
with Bubbles

Equipment needed:
- liquid detergent   - bowl of water
- eggbeater or electric mixer
- several small bowls or dishes
- powdered cake colorings (not liquid)   - paint brush
- sheets of cardboard

Beat a mixture of detergent and water in a bowl until it becomes a very thick foam. Pour the foam into the small bowls, add a different colored powder to each bowl, and mix well.

Now, with your fingers or the paint brush, make patterns or pictures on the sheets of cardboard, using the colored foam as paint. When the paintings are dry, they will have a crusty surface like sandpaper. Why?

After the water and some of the detergent evaporate, the powder and some detergent are left behind. The dry powder keeps the shape of the tiny foam bubbles, many of them broken. The little dry bubbles and broken edges feel rough to the touch.

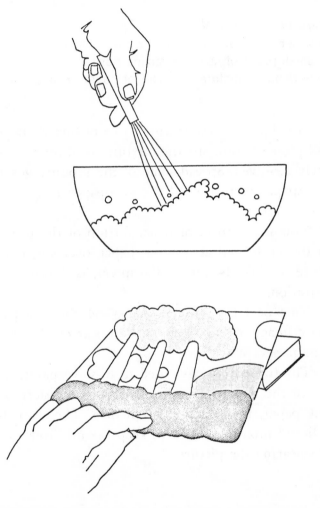

Paintings made with detergent foam, colored by powdered cake colorings, feel crusty after they are dry.

Copying a Newspaper Picture

Equipment needed:
- water   ▪ turpentine or other paint thinner
- small piece of soap   ▪ mixing bowl
- newspaper picture   ▪ sheet of paper   ▪ large spoon

Mix 4 parts of water with 1 part of turpentine and the small piece of soap. Stir the mixture until the soap is completely dissolved. Spread some of the mixture gently over a newspaper picture until the newspaper is evenly moistened.

Now put a sheet of paper on top of the picture and rub the entire surface of the paper over the newspaper picture with the bottom of the spoon, as shown in the top illustration.

Now peel the sheet of paper off the newspaper. A reverse image of the picture will appear on the underside of the paper. Why?

The turpentine, also used to dissolve paint, dissolves some of the newspaper ink. The ink is transferred to the blank paper, reproducing the picture. The soap thickens the liquid mixture and helps keep the ink from spreading and smearing the picture.

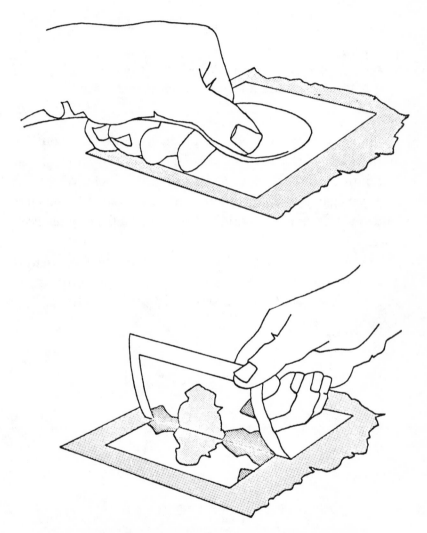

A special liquid mixture, spread between a blank sheet of paper and a newspaper picture, produces a copy of the picture on the paper.

Pushing an Egg
into a Bottle

Equipment needed:
- hardboiled egg ▪ small bowl ▪ vinegar
- bottle with a neck a little narrower than the egg

Soak the hardboiled egg in the vinegar for about 24 hours. The shell will soften so that you can squeeze the egg into the bottle without damaging the shell. Why?

Vinegar is a mild acid that dissolves part of the shell and replaces it with liquid so that the shell becomes flexible.

Vinegar is used to pickle many foods such as cucumbers and tomatoes. Although they are not as hard as an eggshell, they also become softened by the pickling.

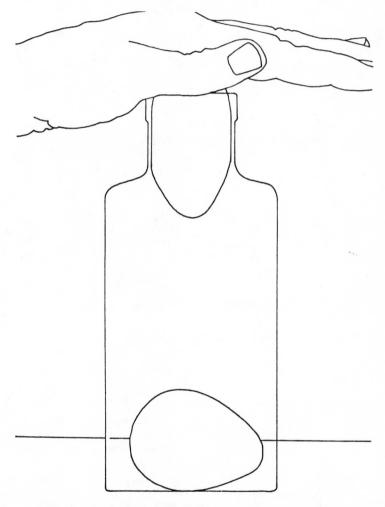

After its shell is softened by a long soak in vinegar, a hardboiled egg can be pushed into a narrow bottle neck without breaking.

viewing
and drawing

Drawing a
Perfect Ellipse

Equipment needed:
- flat wooden board
- two long bulletin board tacks (pushpins)
- sheet of typing paper ▪ scissors ▪ strong thread
- pencil or ballpoint pen

Lay the paper on the board and push in the tacks *a* and *b* about 5 inches apart in the middle of the paper— along the length, not along the width. Push the tacks in just enough so that they will not pull out easily. Place them equal distances from the edges of the paper. Cut the thread 15 inches long and tie the ends together. Put the loop of thread around the tacks.

Now pull the loop tight against the tacks with the point of the pen or pencil and press the point against the paper. If you move the point around the tacks, always pulling against the thread to keep it tight, you will draw a perfect ellipse. Why?

When you follow the instructions, you are obeying the mathematical definition of an ellipse. Each point, *a* or *b*, is called a focus of the ellipse. By definition, if you add the distance between any point on the ellipse and point *a* to the distance between the same point and point *b*, you always get the same number. The sum of the two lengths in this experiment is just the total length of the thread minus the straight part between *a* and *b*. Since this is always the same, no matter where the pen or pencil is, the pen or pencil point draws an ellipse.

Placing the tacks farther apart, or shortening the thread, makes the ellipse more cigar-shaped. Placing the tacks closer together makes the ellipse rounder. If you use only one tack (imagine both tacks at the same spot, with *a* lying on top of *b*), you will draw a perfect circle. A circle is really a special kind of ellipse in which the two focus points are at the same place.

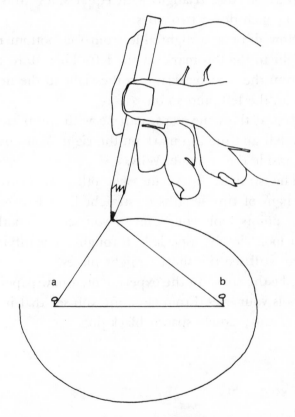

Moving a pencil while pulling a loop of thread tight against two tacks draws a perfect ellipse.

Making a Curve
by Drawing Only Straight Lines

Equipment needed:
- pencil ▪ ruler ▪ large sheet of paper

Draw a large triangle with equal sides. Make marks every ¼ inch down two sides.

Now draw a straight line from the bottom mark on the right to the top mark on the left. Then draw a second line from the next mark up on the right to the next mark down on the left, and so on.

Repeat the same steps, starting with the bottom mark on the left and the top mark on the right. You now have a curve inside the triangle. Why?

The straight lines cut each other at many places. The chain of tiny lengths of straight lines between inter-section points fools your eyes so you see a smooth curve. If you look closely, especially through a magnifying glass, you can still see the short straight pieces.

Like the curve in the experiment, a newspaper picture also fools your eyes. From close up, you see that it is really made of tiny, closely spaced black dots.

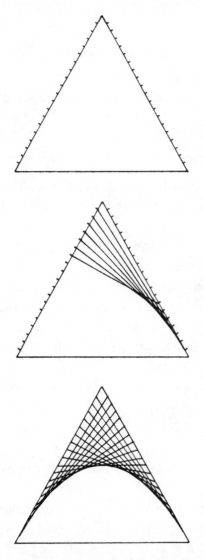

When you draw many straight lines that intersect each other at points close together, the chain of tiny straight lengths looks like a smooth curve.

Finding the Center
of Africa

Trace the outline of Africa. Place the tracing on the cardboard and cut around it. Push the pin through the cardboard near an edge at point *a,* as shown in the top illustration, and pin the cardboard to a wall. Make sure the cardboard swings freely around the pin.

Tie one end of the thread around the small weight and loop the other end over the pin. Now, without disturbing the hanging weight and the thread, draw a line on the cardboard exactly under the thread. Remove the pin.

Now pin the cardboard to the wall through another point *b,* about one-quarter of the way around the edge of the cardboard, as shown in the second illustration. Hang the weight from the pin as before, and again draw a line under the thread.

The point *d* where the two lines cross is the center of the map of Africa. Why?

When the cardboard is allowed to dangle freely from the pin, it adjusts itself so the left half balances the right half. The pull of gravity on the hanging weight makes the thread follow the dividing line about which the two halves are balanced. Repeating this step at another point provides a second line about which the map of Africa is balanced.

146

The crossover point *d* of the two lines is therefore the center point of the map. When the map is pinned to the wall at any other point, like *c* in the third illustration, and the weight is hung from the pin, then the string will always pass through the same center point.

Such a point is called the center of gravity. The center of gravity is important in many practical ways. For example, when a mechanic adds weights to an automobile wheel to balance the tire, he is adjusting the center of gravity to the center of the wheel.

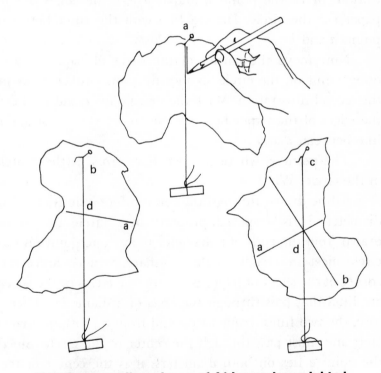

Hanging a cardboard map of Africa and a weighted thread from a pin helps to locate the center of the irregular shape.

Finding the Center
of a Circle

Put the square corner of the sheet of paper anywhere on the boundary of the circle, as shown in the first illustration. Mark the points *a* and *b* where the edges of the paper cut the circle. Draw a line with the ruler between points *a* and *b*.

Now move the corner of the sheet of paper to any other point on the circle, as shown by the broken lines in the second illustration. Mark the new points *c* and *d* where the edges of the paper now cut the circle. Draw a straight line between *c* and *d*.

The point *e,* where the two lines cross, is the center of the circle. Why?

A line drawn through the center of a circle is called a diameter. It is a special property of a circle that lines drawn from the ends of a diameter to any one point on the circle meet in a right angle (a square corner). Since the corner of the sheet of paper is square, the edges of the sheet are lines that pass through the ends of a diameter. Therefore, the two lines (from *a* to *b* and from *c* to *d*) are diameters and must pass through the center of the circle. Since the point *e* lies on both diameters, it is the center of the circle.

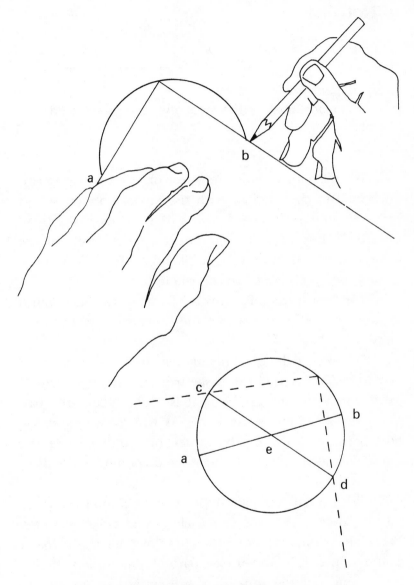

Using the square corner of a sheet of paper allows you to locate the center of the circle.

Proving that the Earth Is Not Flat

Ask a friend to hold the sheet of wood or cardboard horizontal at chest height with the model boat on top of his or her end of the board. You walk from across the room toward the boat. The boat gradually appears larger as you come closer, as shown in the left-hand column of the illustration, but you always see the whole boat.

If the earth were flat, this is how an ocean liner would look to you as it approached the shore from many miles away.

Now repeat the experiment, but this time your friend holds the boat behind the basketball (still at chest height) so that only the top of the boat is visible above the ball when you stand across the room. As you come nearer, the boat not only seems to grow larger, but you see more and more of it, as shown in the right-hand column of the illustration. Why?

This is what you actually see when a ship first appears on the horizon and then comes closer and closer to shore. When the ship is far away, the curve of the earth is like a round-topped hill (the upper part of the basketball) between you and the ship. The hilltop is like the earth's horizon. The horizon looks straight only because the earth is so much bigger than a basketball. As the ship steams toward you, it rises up over the curve of the earth and becomes more and more visible. You do not see the whole ship until it reaches the horizon (the top of the "hill"). Then the ship continues down your side of the earth's curve to your position on the shore. The ocean near you seems flat for the same reason that the horizon seems straight—the ball of the earth is so much bigger than the basketball.

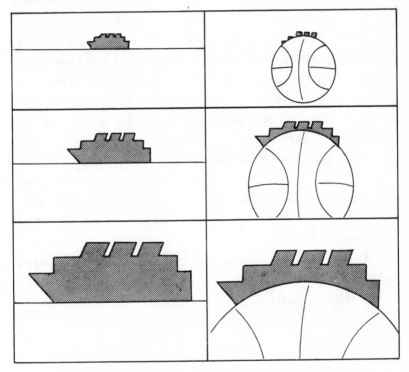

As you come nearer to a small model boat, it looks like the boat in the left-hand column if it is on a flat surface. It looks like the boat in the right-hand column if the surface is round like that of a basketball.

Turning an Arrow Around
without Touching It

Equipment needed:
- piece of cardboard   ▪ pencil or pen
- plain, round, clear glass of water (no ripples, or flat sides) about 3 inches in diameter

---

Draw an arrow on the cardboard as shown in the top illustration. Stand the glass of water about 6 inches in front of the arrow and look at the arrow through the glass from about 6 inches in front of the glass. The arrow you see will point in the opposite direction from the arrow on the cardboard. Why?

The glass and water act as a lens does in a camera. Light rays from the arrow are bent when they pass through the curved surfaces of the glass of water. They cross over each other before they reach your eyes so you see the arrow backward.

Lenses in a microscope also bend light rays and spread them out wide so that the object you look at appears much bigger than it really is.

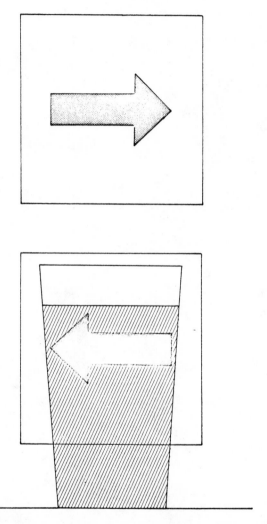

When seen through a round glass of water, the arrow points in the opposite direction.

electricity
and magnetism

Moving Two Balloons Apart without Touching Them

Blow up the balloons to equal size. Rub each of them with the cloth several times in one direction only.

Now quickly tie each end of the string to a balloon, touching the balloons as little as possible. Grasp the center of the string and hold the balloons away from your body. The balloons will move away from each other. Why?

There are two kinds of electric charge, called positive and negative. Charges of the same kind push each other away. Charges of opposite kinds attract each other. Rubbing the balloons with wool charges them both with negative electricity. Since the balloons both carry the same kind of charge, they move away from each other.

If you electrically charge yourself by rubbing your feet across a woolen carpet, the hairs on your head will stand out. They all have the same kind of electric charge, so they move away from each other like the balloons in the experiment.

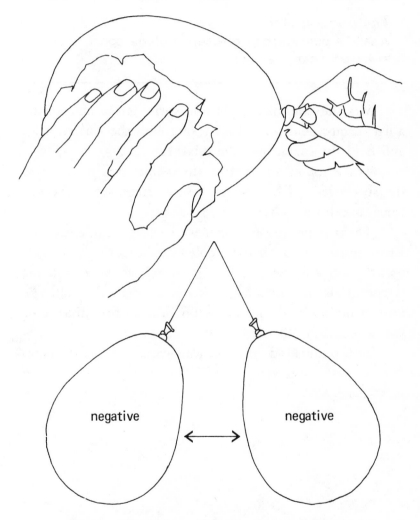

Balloons rubbed with a woolen cloth move away from each other.

Separating Salt from Pepper without Touching Them

Equipment needed:
- salt ▪ pepper (finely ground) ▪ plastic spoon
- woolen cloth

Mix the salt and pepper and sprinkle the mixture on a flat surface. Rub the plastic spoon with the woolen cloth and hold the spoon over the mixture.

As the spoon comes close, the pepper will jump up to the spoon and stick to it for a short time, while the salt remains behind. Why?

The plastic spoon becomes charged with electricity when you rub it with the woolen cloth. Both the pepper and the salt are attracted by the electrified spoon, but the pepper rises first because it is lighter than the salt. (Be careful not to hold the spoon too close to the mixture or you will pick up the salt as well.)

The same kind of electrical attraction often causes dust to gather on your television screen and on charged parts inside the set.

A plastic spoon, after being rubbed with a woolen cloth, attracts lightweight pepper before heavier salt.

Bending a Stream of Water without Touching It

Equipment needed:
- plastic comb
- thin, steady stream of water from a faucet

Pass the comb quickly through your hair several times. Your hair must be dry. Now bring the edge of the comb near the falling stream of water, taking care not to wet the comb.

When the comb is about ¾ inch away, the stream of water will start to bend toward the comb. Why?

Rubbing the comb against your hair charges the comb with static electricity. Even though the water is not electrically charged by itself, a charge is induced on the surface of the water near the comb. Since the electricity on the comb and the induced electricity on the water surface are of opposite kinds, they attract each other. The attraction draws the water toward the comb.

Rubbing builds up electricity on many familiar things in your everyday life. For example, bedsheets and other laundry items tumbled in a clothes dryer become so electrically charged that they cling together and must be peeled apart.

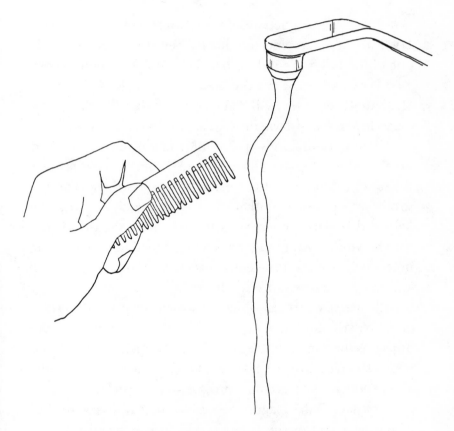

A plastic comb passed through your hair several
times attracts a stream of water.

Balancing Matchboxes
with Magnets

Equipment needed:

- two small matchboxes
- two bar magnets that fit into the matchboxes

Place the magnets inside the boxes and close the covers. Put the boxes end to end. If they stick together, turn one of the boxes around so that they push each other apart.

When you now tilt the boxes upward and bring them close together, they will balance in midair like an open draw bridge, as shown in the top illustration. Why?

The opposite ends (poles) of a bar magnet are called north and south. The north (*n*) end of one magnet attracts the south (*s*) end of another magnet. Two north ends or two south ends push each other away. The matchboxes can be balanced because gravity tries to make the raised ends fall, which would bring the ends closer together, while the magnets inside push the ends apart. At balance, the gravity and magnet forces cancel each other.

A magnet attracts objects containing iron or certain other metals—for example, a steel needle—even though the object is not itself a magnet. This happens because the magnet magnetizes the needle when the needle is brought close enough. The needle is magnetized in such a way that the part facing the magnet becomes a pole opposite to that of the near end of the magnet. Therefore, the two attract each other. Even after the needle is pulled away from the magnet, some magnetism remains in the needle. The needle is then a weak magnet that can attract other needles.

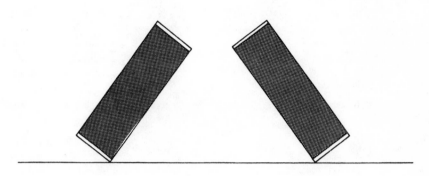

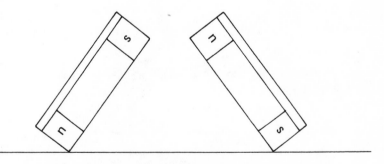

Two matchboxes with magnets inside are balanced with their ends raised in midair.